THE TIGHTER HOUSE

By Charlie Wing

With John Lyons and the Staff
of Cornerstones, Brunswick, Maine

Edited by Larry Stains,
Executive Editor,
New Shelter® Magazine

Rodale Press Emmaus, Pa.

Most of the information presented here originally appeared as a series of articles entitled "The Tighter House," and as other articles in *New Shelter*. Special thanks to: Joe Carter, Jim Eldon, Michael Lafavore, Frederic S. Langa and William Shurcliff.

Printed in the United States of America on recycled paper containing a high percentage of de-inked fiber.

Book design by Merole Berger

Library of Congress Cataloging in Publication Data

Wing, Charles, 1939-
   The tighter house.

   "'Most of the information presented here originally appeared . . . in New shelter."
   Includes index.
   1. Dwellings—Insulation. 2. Dwellings—Heating and ventilation. I. Lyons, John (John J.) II. Stains, Larry. III. Cornerstones (Firm) IV. Title.
TH1715.W54   643'.7   81-17694
ISBN 0-87857-378-X   paperback AACR2

2 4 6 8 10 9 7 5 3 1   paperback

Contents

1 Home Sweet Leaky Bucket........................1

2 Basic Tools: Insulation and Ventilation.............4

3 From the Ground Up: Firmly Insulated
Foundations.....................................15

4 Insulating Your Walls and Floors.................24

5 Insulating the Attic.............................32

6 Doors: When You Shouldn't Bother...............40

7 Windows: Many Solutions to a
Major Problem..................................44

8 Make Your House Float: Caulking and
Weatherstripping...............................51

9 The Heat Leak Hit List.........................60

10 Hot Water: Waste Not, Sacrifice Not.............66

11 Heating Systems: Getting More Degree Days
per Gallon......................................72

12 Priorities: Being Your Own Energy Manager........80

Index...91

Chapter 1

Home Sweet
Leaky Bucket

THE TIGHTER HOUSE

Our understanding of heat and its workings is the single most important element in our determination to reduce our energy bills. Heat is defined as energy that flows between bodies as a result of a temperature difference. If you have a warm object next to a cool object, heat flows from the warmer to the cooler until both reach the same temperature. It is important to understand that heat flows in only one direction: hot to cold.

Before we can measure heat and develop our plan to control it, we must first differentiate between heat and temperature. Temperature, as we all know, can be measured directly with a thermometer. Heat, on the other hand, can be measured only indirectly, by calculation. Temperature is a measure of how hot something is. How much heat an object contains depends on its composition and size as well as its temperature. Heat and temperature differ in another very important way: heat can be produced, purchased, and put to work; temperature cannot. You do not buy temperature.

The basic unit used in measuring heat is the British thermal unit (Btu). The Btu is the quantity of heat required to raise the temperature of one pound of water one degree Fahrenheit. The Btu may not be very exciting to you, but it is a very important unit, by which all of the heat energy within your residence is measured. Burning one gallon of oil at 70 percent efficiency produces approximately 100,000 Btu's.

Here are some other ways to measure heat, and the lack of it. You'll find these terms used regularly because they are the key variables that determine the energy needs of your home.

- Degree Day (DD): a unit indicating the difference in degrees Fahrenheit between the average daily temperature and the accepted standard of 65°F.
- Annual Degree Days: the total number of degree days added together over all the days of the heating season. Caribou, Maine, usually has about 10,000 annual degree days, while Atlanta, Georgia, normally has only about 3,000 degree days annually.
- Design Minimum Temperature: a temperature taken from weather

data indicating the lowest expected temperature for a given location.
• Design Heat Load: the total energy needed per hour to heat a building when the outdoor temperature is at the Design Minimum Temperature.
• Heat Loss: the rate of heat flow out of a building per unit of time, usually one hour.
• Heat Gain: the rate of heat flow into a building per unit of time.

Annual degree days and design temperature are functions of the climate and are consistent for all homes in a general location. They are indications of the length and severity of the heating season. Design heat load, heat loss, and heat gain are specific measures for each individual building. Together with degree days and design temperature, they determine the amount of energy necessary to maintain comfort within a specific home.

A BUCKET OF HEAT

We can all relate to the frustration of attempting to transport water from one location to another with a bucket that has a hole in it. The simple analogy of a leaky bucket (see Illus. 1.1) can help you understand the heat losses and heat gains of your residence. The bucket represents the house; the fluid entering the bucket represents the heat introduced into the house to maintain a comfortable inside climate; and the fluid escaping through the hole represents the total heat loss of the building. Developing the analogy further, the faucet represents auxiliary heat,

Illustration 1.1
The Bucket Analogy

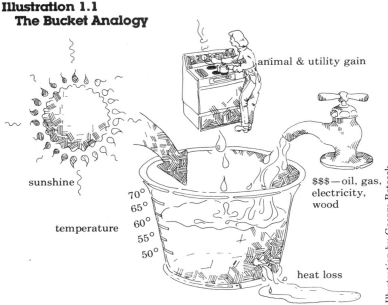

animal & utility gain

sunshine

temperature

70°
65°
60°
55°
50°

$$$—oil, gas, electricity, wood

heat loss

Illustration by George Retseck

2

usually in the form of purchased fuels (oil, gas, wood, etc.). Most of your heat gain probably comes in this "store-bought" form. The person at the stove represents what is referred to as animal gain. That's right—our warm bodies crank out more than 300 Btu's per hour. The stove represents heat gained as a by-product of other household operations, commonly referred to as utility gains. The sun obviously represents solar gain, including the heat of the sun streaming in through ordinary windows.

Now, imagine that our bucket is sitting in a larger pool of water and that the level of water in the larger pool comes up exactly to the level of water in the bucket. No leaks, right? Of course not. But if the level of the outside water drops, then the bucket also loses water—that is, heat. Hence the temperature levels along the left-hand edge of the bucket. The lowering of the outside water level (outside temperature) or the raising of the level inside the bucket (indoor temperature) increases the pressure and rate of flow of heat from inside to outside, or from hot to cold.

The same analogy works for homes that have summer cooling needs. If the outside water level—temperature—rises *above* the water level of the bucket, water—heat—will rush *into* a leaky bucket.

In a nutshell, that's why our houses lose heat in the winter and get hot in the summer. The question is, what can we do about it? Our objective, remember, is to maintain the level of water in the bucket using a minimum of store-bought fuel. We do not have control over the exterior temperature, nor do we have substantial control over the animal and utility gains that are results of everyday living. Here are the critical variables that determine the amount of auxiliary energy we must purchase:

- The first step to take is to plug the hole. Reducing the size of the hole will restrict the flow of heat to the outside, which in turn will reduce the amount of auxiliary heat required to maintain the comfort level. Heat leaks in houses should be plugged with insulation, caulking, and weatherstripping.
- Solar gain has the potential to contribute a greater percentage to the total energy load. We can lower the energy bill by adding south-facing glazing, if we insulate that glazing at night. Solar energy also can be used to heat our tap water, which is second only to space heating in consumption of home energy.
- Finally, a shameful amount of the auxiliary heat we purchase never makes it into the bucket, but dribbles uselessly overboard! There are a variety of improvements we can make to our heating systems themselves that will result in more efficient use of our store-bought energy.

 # Basic Tools:
Insulation and
Ventilation

THE TIGHTER HOUSE

As we found in Chapter 1, heat flow is the transfer of energy between objects when there is a difference in temperature. In the leaky bucket analogy, water flowed out of the bucket when there was a difference in water level between the inside and outside of the bucket. Heat flow is different from water flow in that it happens in not one, but three ways. Understanding the three heat flow processes will help define methods of plugging the holes in your bucket.

Conduction is the process of heat transfer whereby heat moves directly through a material. An example often used to display this phenomenon is a silver spoon in a cup of hot coffee. Heat from the coffee is transferred through the spoon by conduction. It is important to note that different materials have differing abilities to transfer energy internally. Those that transfer energy readily (such as silver, copper, and aluminum) are called "thermal conductors." Materials that conduct poorly (such as wool and air) are called "thermal insulators." So it follows: If you are attempting to reduce heat loss, you are interested in using thermal insulators to restrict heat flow.

Convection is a transfer of energy by the motion of fluids, primarily air, due to temperature-caused density differences within the fluid. Hot air rises and is replaced by cooler air; the shuffle creates convection currents. Infiltration is a special type of convection that is a major source of heat loss from most buildings. The draft you feel as you sit beside a window is an example of infiltration. Infiltration occurs most often through cracks around windows and doors and through uninsulated walls. Infiltration is often easy to locate and relatively easy and inexpensive to eliminate.

The third heat loss process, **radiation,** is the most difficult to understand and calculate. But it's also the least influential, so we will simply define it: Radiation is the process of heat transfer through space; it needs no medium, such as air or a solid material, to be transferred.

Within your residence, these three movements of heat occur most often simultaneously and in conjunction with one another. A wood stove is a good example of this. Heat is conducted from the inside of the

4

firebox to the outside, where it is radiated to other objects in the room. There it heats up the adjacent air, making it less dense, causing it to rise and be replaced. Voilà, a convection current.

R-VALUE

We now know how heat flows, and we have identified our major objective as reducing heat loss, or plugging the hole. What is needed in order to compare heat losses and the relative value of conservation measures is a common unit. R-value—thermal resistance—is a measure of the relative ability of a material to slow down or resist heat flow. Thermal resistance relates to conduction. Each material in the wall of your home has an R-value. Added together, they give you an overall value. The total resistance, as we have already concluded, is directly related to heat loss. The R-value of materials, such as "insulation," helps to compare them and assess the economic value of your actions.

The rate at which heat flows through a material is inversely proportional to the total R-value of the material. That is, the greater the R-value, the smaller the heat flow. For example, the R-value of an ordinary single-glazed window is about 1; the R-value of a double-glazed window about 2 (generally each layer of glazing increases the total R-value by 1). Because of the inverse relationship between heat loss and R, the double-glazed window loses only half as much heat. A six-inch-thick batt of fiberglass labeled R-19 would lose only 1/19 (about 5 percent) as much heat as the single-glazed window of the same area.

R-value is thus a very telling number. The Federal Trade Commission now requires that insulation manufacturers, installers, and sellers label or otherwise inform the customer of the R-value of insulation products.

Strange as it may seem, there are few better insulators than ordinary dead air. The problem is, dead (not moving) air is not so ordinary! Whenever a temperature difference exists, air tries to eliminate the difference by transferring heat from warm to cool by convection. An extraordinary fact is that most commercial insulations are nothing more than low-density materials that stop air from moving. These insulations fall into four broad categories: (1) blanket and batt, (2) loose-fill and loose-pour, (3) rigid foam boards, and (4) blown-in.

Table 2.1 summarizes the more salient characteristics of the common insulations used in the home. The "combustible" characteristic of cellulose and all of the foams is a rating established by a laboratory procedure and does not necessarily reflect a degree of fire danger in the home. In fact, properly treated and installed, none of the "combustible" insulations pose any more fire danger than your household furnishings.

VAPOR BARRIERS AND VENTILATION: THE PROBLEM

Insulation is the cork we stuff into the holes in our heat loss bucket. Unfortunately, insulation cannot be considered without moisture. Moisture can get trapped inside insulation. When it does, the insulation

5

BASIC TOOLS
Table 2.1   Insulation Table

Form and Type		R-Value per Inch	Cost	Characteristics
BLANKET AND BATT				
Fiberglass (spun glass fibers)		3.2	low	noncombustible except for facing difficult with irregular framing
Rock wool (expanded slag)		3.4	low	noncombustible except for facing difficult with irregular framing
LOOSE-FILL[1]				
Fiberglass	attic	2.2	low	noncombustible
	walls	3.8		good in irregular spaces
Rock wool	attic	2.9	med.	noncombustible
	walls	2.9		good in irregular spaces
Cellulose (paper	attic	3.7	low	combustible—specify "Class I, noncorrosive"
fiber)	walls	3.3		can be damaged by water
Perlite (glass beads)	attic	2.5	high	noncombustible
	walls	3.7		expensive
Vermiculite (expanded mica)	attic	2.4	high	noncombustible
	walls	3.0		expensive
RIGID FOAM BOARDS[2]				
Molded polystyrene (bead board)		4.0	med.	combustible permeable—do not use below grade maximum temperature 165°F
Extruded polystyrene (Styrofoam)		5.0	high	combustible impermeable—best below grade maximum temperature 165°F
Polyurethane Polyisocyanurate (Thermax, etc.)		6.0 7.4	high	combustible good for walls and roofs outside maximum temperature 250°F
BLOWN-IN FOAM				
Urea-formaldehyde		4.2	med.	combustible use in closed wall cavities only not damaged by water may exude formaldehyde odor and shrink more than 3% if improperly applied

Data taken from *An Assessment of Thermal Insulation Materials and Systems for Building Applications,* Brookhaven National Laboratory, June, 1978, GPO Stock No. 061-000-00094-1. Source: Cornerstone Energy Audit.

1. All loose-fill insulation must be installed at manufacturer's recommended densities as shown on bag to insure proper performance.
2. All rigid foams are combustible and must be covered with ½-inch drywall or equivalent 15-minute fire-rated material when used on interior.

cannot operate as efficiently, and more heat loss occurs. But the heat loss is not as great a concern as "house loss." Moisture, when trapped for prolonged periods under the proper conditions, can cause sheathing and roofing boards—and in severe situations, framing members—to decay. In order to have a tightly insulated house that won't slowly rot out from under us, we have to understand how water vapor in the air becomes moisture, and then how to keep moisture out of our attic and walls.

Water occurs in three states: solid (ice), liquid (water), and gaseous (water vapor). We are familiar with the solid and the liquid states of water but not so familiar with the gaseous state. That's because, like heat, it can't be seen. You may think you've seen water vapor rising up from a whistling tea kettle, but in reality what you see is condensed water vapor, a liquid. Water vapor is a natural component of air, much like nitrogen and oxygen. There is one important difference: Air can tolerate only a limited amount of water vapor. Beyond this upper threshold, called the dew point, the vapor condenses and is forced out as liquid. This is where the problem begins. The amount of water vapor that air can hold depends on the air temperature and is measured by relative humidity. Relative humidity is the actual amount of water vapor in the air compared to the maximum amount possible at a given temperature.

Warmer air can hold more moisture than colder air. The warm air within our homes is able to hold the moisture normally produced within a household. But as this air moves toward the exterior of the house (remember, heat always flows in one direction: hot to cold), the temperature drops, and the air becomes less able to hold water vapor. Upon reaching the dew point, the water vapor condenses.

The point where this condensation occurs is critical. In an uninsulated wood frame home, for example, the temperature drop occurs rapidly within the wall cavity, and the condensation forms, either as dew or frost, on the first cold surface it encounters, usually the inside of the exterior sheathing. In the typical uninsulated wall cavity, there is ample ventilation to evaporate and transport the moisture to the outside, avoiding the problem of standing water and decaying wood.

In the insulated wall, more condensation occurs because the exterior sheathing is now colder than before. The condensation builds up in the insulation as a solid layer of ice. Upon melting, the insulation absorbs the moisture and inhibits the circulation of air that previously carried the moisture away.

The seriousness of this problem depends upon where you live. Generally speaking, the warmer and drier your climate, the less you have to worry about condensation.

Moisture trapped within insulation can do two harmful things. As we said, it can slightly decrease the efficiency of the insulation. To the extent that fibrous insulation with its trapped air pockets is replaced by solid ice, the amount of insulation is decreased. Potentially more damaging is the creation of a favorable environment for wood decay because the insulation retards evaporation of moisture. The wood-eating fungus commonly referred to as "dry rot" is promoted by warm temperatures (above 50°F), moisture, and the absence of light. If the moisture that collects in the insulated cavity during the winter remains until temperatures reach 50°F, you may have a problem.

THE SOLUTIONS

Now let's find out how to control moisture and condensation. As long as the moisture stays in the air and keeps moving, no harm is done. Our objectives are to keep the moisture in the air and to avoid trapping moisture within the building cavities. The successful control of moisture involves three important areas.

First, we can reduce the amount of water vapor produced in the home. Reducing the amount of water vapor (and thus the relative humidity) within the home reduces the chances for saturation and condensation by lowering the temperature at which condensation occurs. Lower relative humidity results in the dew point occurring at a lower temperature, so condensation happens less frequently.

The second weapon we have to stop water vapor before it reaches the insulation is a vapor barrier. A vapor barrier is a material or coating that successfully inhibits the passage of water vapor (see Table 2.2). A vapor barrier applied to the inside surface of insulation prevents moisture from entering the cavity.

The last weapon in our three-pronged attack against moisture is ventilation. The introduction of outside air helps to remove moisture from the cavity. The exterior skin of a structure must be allowed to "breathe" in order to eliminate unwanted and potentially damaging moisture.

To summarize, our three-pronged solution follows the path of the problem:

1. Produce less water vapor in the house
2. Block vapor that is unavoidable with a vapor barrier
3. Ventilate moisture that penetrates the vapor barrier

Most people think of winter as a miserably dry time inside their homes. Nasal passages feel parched and the skin dry. Some people boil water or set pans of water atop radiators to moisten arid air. Before the age of tight houses these remedies were necessary because the infiltrating

Table 2.2 Permeance of Vapor Barriers

Material	Permeance
Aluminum foil	0.0
Polyethylene film 4-6 mil	0.08
Aluminum paints	0.5
1 coat latex vapor barrier paint	0.6
Vinyl wallpaper	1.0
2 coats oil paint on plaster	2.0
3 coats latex on wood	10.0
Ordinary wallpaper	20.0

Note: Permeance values less than or equal to 1 are considered adequate vapor barriers.

Table 2.3 How Moisture Is Added to Air

Activity	Lbs. of Moisture
Washing clothes, per week	4.0
Drying clothes on indoor line, per week	26.0
Cooking and dishwashing, per week	35.0
Each shower	0.5
Each tub bath	0.2
Normal respiration and skin evaporation, per person per 24-hour day	2.9

Source: *Keeping the Heat In*, Energy, Mines and Resources, Canada, Office of Energy Conservation, August, 1976, p. 17.

air was absorbing the water vapor within the building and carrying it to the outside.

In the tightened-up house, dry air is no longer a problem. Air exchange rates have been reduced, and the considerable amount of moisture that is added to inside air by normal daily activities (see Table 2.3) has an opportunity to build up. So you are faced with a new problem: too much moisture. A few logical adjustments are in order: keep lids on your pots when cooking, don't dry laundry inside, and perhaps open the bathroom window a crack after showering. A hose should be used to expel your dryer's very moist air to the outside. Do not attempt to dry large quantities of firewood inside. A cord of green wood contains as much as 2,000 pounds of water.

In addition to these sources, there is one major moisture source to watch out for: the area beneath your home. Whether it be a full basement or a perimeter foundation with a crawl space, moisture from this area must be eliminated or prevented from entering the living area. Most often a wet basement is caused by inadequate drainage around the exterior of the foundation. There are two ways to solve this problem: (1) remove the water as it enters, or (2) prevent the water from entering in the first place.

The first alternative is obviously only a partial cure, but as you might expect, it's cheaper. Dig a trough (difficult in concrete) at the lowest point in the floor, or around the outside edge where the moisture can be

Table 2.4 Crawl Space Ventilation

Crawl Space	Ratio of Total Net Ventilating Area to Floor Area	Minimum Number of Ventilators
Without vapor barrier	1/150	4
With vapor barrier	1/1,500	2

Source: *Condensation Problems In Your House: Prevention and Solution*, USDA Forest Service, September, 1974, p. 12.
Note: Foundation vents should be distributed to provide the best air movement.

collected, and install a sump pump to transport the water to the outside. A second, and more effective, cure is exterior excavation and the installation of a drainage system, consisting of gravel and drainage tiles that carry the water away from the foundation. A third cure is to apply a water repellent to the inside of basement walls. None of these approaches is easy or cheap, but the elimination of moisture is critical to both internal comfort and the structural integrity of your home.

Water vapor from a crawl space beneath a home also should be prevented from entering the home and causing damage. To accomplish this, place a sheet of 6-mil-thick polyethylene over the soil to act as a vapor barrier. The air space above should then be vented. Table 2.4 offers suggestions for adequate ventilation. Plug the ventilation openings in winter with movable or temporary insulation to reduce heat loss. If the crawl space is used for storage, the vapor barrier can be protected by covering it with a two-inch-thick layer of sand.

VAPOR BARRIERS

A very effective way of controlling the moisture problem in the building envelope is the use of vapor barriers. A vapor barrier is any material which does not allow water vapor to pass through it. A good vapor barrier also serves as a good wind or air barrier, helping you control infiltration. Vapor barriers should be placed on the warm side of all insulated surfaces. Fiberglass insulation can be purchased with a foil-backed, vapor barrier facing. Although this foil theoretically serves as a vapor barrier, a continuous layer of 6-mil (1 mil = 0.001 inch) polyethylene is often applied over the foil because the polyethylene can cover an entire wall without gaps and holes. Polyethylene should be used wherever the situation permits. It is the best vapor barrier for the price. Either 4- or 6-mil can do the job, but 6-mil is recommended because it is less likely to tear during installation.

Illustration 2.1

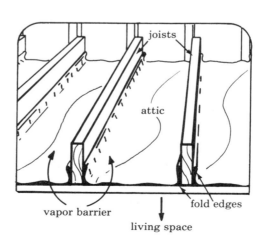

joists

attic

vapor barrier

fold edges

living space

Illustration by George Retseck

Polyethylene is available at most hardware stores and building supply outlets. "Poly," as it is sometimes called, is purchased in rolls of 8-, 10-, 12-, 16-, and 20-foot widths and lengths of 25, 50, 100, and 200 feet. The larger rolls are cheaper per square foot of material, so it is wise to figure your whole job from the start.

For walls and rafters, you should stretch a continuous sheet of polyethylene across the inside face of the studs or rafters after the insulation is in place and staple it into place. Staple across the top first, then down one side or corner. Continue stapling across the sheet, remembering to keep the film taut and flat so you do not create problems with the application of the interior finish. Hammer down any raised staples. Cut away the barrier around electrical boxes and outlets. Duct tape, available at a local hardware store, should be used to seal around outlets and along all seams that do not overlap another sheet of vapor barrier at a rafter or stud. Where sheets of vapor barrier meet, try to achieve an overlap of 12 inches.

If insulation is to be placed between the floor joists of the attic, the polyethylene should be cut in strips and stapled to the sides of the joists (see Illus. 2.1). Cut the strips four inches wider than the distance between the joists and fold the edges over to insure a tight seal. Place the insulation on top of the vapor barrier. Be careful not to block attic ventilation around the soffit area of the structure.

If you insulate the floor beneath your living space, the vapor barrier should go on first, fitting up under the bottom of the subflooring between the floor joists, as in the attic. The strips should be cut with the same additional four inches so they can be folded over and stapled. Then fit strips of insulation between the floor joists.

Illustration 2.2 shows the vapor barrier on the inside of the insulation repelling water vapor from the interior of the building envelope. This simplified illustration represents the ideal vapor barrier situation. The entire living area is enclosed within the protective envelope of the vapor barrier.

Illustration 2.2

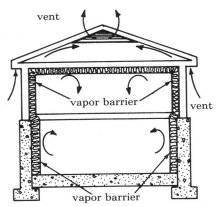

Illustration by George Retseck

In new construction this is easy to do, but it is very difficult to wrap an existing house in a protective plastic bag of polyethylene. As a matter of fact, it is next to impossible unless you are gutting the building (hardly justifiable purely for the sake of insulating). Table 2.2 offers some alternative vapor barriers. Vinyl wallpaper, although expensive, is offered in a variety of patterns and colors. Other acceptable alternatives are oil-based paint and some new lines of latex vapor barrier paints that are very effective in stopping moisture.

Is a totally effective vapor barrier absolutely essential? No. Reduced vapor production, vapor barriers, and ventilation all work together toward the same goal: reducing moisture within the dark cavities of your home's shell. If it is difficult or excessively expensive to combat moisture from all three directions, it is important to perform tasks that can be done. If a vapor barrier is impossible, moisture production in that area should be kept to a minimum and ventilation provided.

Most exterior walls breathe very well. The exceptions, however, are walls covered with aluminum or vinyl siding over unperforated foil paper. In this situation, ventilation from the interior of the house to the outside cannot occur. Even though some foil papers are perforated with small pinholes, don't take chances. If you have foil, you may have problems after adding insulation. The best solution is to remove rows of siding and either eliminate or slash the foil. Vent plugs once recommended in this situation are usually ineffective in a cavity full of insulation.

The kitchen, bathroom, and laundry area deserve special attention. If a vapor barrier is impossible in these areas, venting becomes more important. The kitchen and the bathroom should be equipped with mechanical venting, and the dryer in the laundry room should have an exhaust hose. As rooms come in need of redecorating or repair, remember to install a vapor barrier.

ATTIC VENTS

The attic is the prime candidate for condensation problems. Most roofing materials are excellent vapor barriers, the exceptions being cedar shingles and tile. Water vapor that collects in the attic space is not able to be carried away as it is in the wall cavity with its naturally breathing exterior. Although vapor reduction and vapor barriers are important, ventilation is the major line of defense against condensation in the roof and attic. Air must be allowed to enter and exit this space and take with it unwanted moisture.

The recommended ventilation of attics is given as a ratio between total vent area (inlet plus outlet) and the total floor area. This ratio is 1:150 for an area without a vapor barrier on the ceiling and 1:300 with an effective vapor barrier. The inlet and outlet areas should be roughly equal in size. Thus, if your attic floor area is 1,500 square feet, and you have installed a vapor barrier, you need a total of 5 square feet of ventilation opening.

The type of venting system depends on the roof style. The object is to exchange the air. One method is to take advantage of the "chimney effect" by introducing air at the lower regions of the roof and exhausting it at the highest point. A common and very effective system is a combination of soffit and ridge vents. Fresh air enters through soffit vents to replace hot attic air, which naturally rises out through vents at the ridge. Various types of soffit vents are pictured in Illustration 2.3. The continuous screened vent is best because of its efficiency and attractiveness.

Gable end venting is another relatively simple method of expelling moist air. Vents are installed toward the peak on both end walls of the roof, allowing air to move through the attic.

Illustration 2.3
Soffit Vents

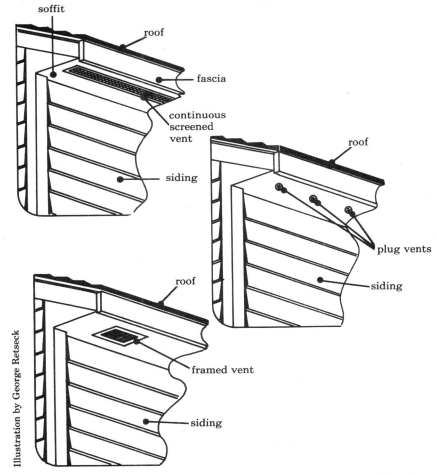

Illustration by George Retseck

If your attic cannot accommodate a natural convection venting system, a mechanical fan can be installed. These roof exhaust fans are thermostatically controlled. The help of a heating/ventilation contractor or a fan salesperson is valuable in determining the proper fan and installation method.

There are a variety of natural ventilation aids not mentioned here. To find one that's right for your situation, the best advice comes from an insulation contractor.

Moisture cannot be neglected in your plans to make your home more energy efficient and economical. Keep in mind that your home is a dynamic system and that measures taken in one area can affect another dimension. The insulation/moisture relationship is one that you, as your own home "energy manager," should understand and work with. Now that the dynamics of moisture are understood, we can get on to plugging the holes.

■■Chapter 3

From the Ground Up: Firmly Insulated Foundations

| THE TIGHTER HOUSE |

Insulating the foundation is too often neglected in the home owner's plan to tighten up his house. The fact is, the foundation can be one of the primary heat loss areas and a major cause of discomfort (cold feet). It may surprise you to learn that the thermal resistance of an exposed concrete wall is less than that of a double-glazed window. That means you're wasting an incredible amount of heat from ground level up to the top of the foundation.

OUTSIDE OR IN?

There is debate over whether to insulate foundation walls internally or externally. Of course, there are specific situations where interior insulation is preferable, such as building on ledge (an underground layer of rock). But if your house has a crawl space or a full basement and you could go either way, exterior insulation is usually considered best for the following reasons:

- Thermal mass. Including basement walls within the insulated envelope increases the thermal mass of the house, thereby helping to store heat, moderate temperature swings, and increase comfort.
- Moisture. Exterior insulation helps to keep moisture out of the basement. (The direct relation between groundwater and heat loss is discussed below.)
- Space. No internal space is sacrificed for the insulation.
- Frost action. External insulation, properly placed, prevents foundation upheaval due to the expansion of the earth as a result of frost.
- Fire safety. Flammable products that might be used for insulation are kept safely on the outside.

But insulating the exterior also has its drawbacks. First and most obvious, owners of existing homes have to dig up the earth along the foundation wall. Second, the rigid foams commonly used must be protected from exposure to sunlight. Finally, exterior installation cannot be done in winter in most areas of the nation. Therefore, don't rule out interior insulation before considering its potential advantages:

15

- The insulation may represent a minor additional cost if you are already planning to finish the basement as a living area.
- The insulation may provide acoustic improvement.
- The work can be done at any time of day or year.

If you do decide to turn to the inside, a strong word of caution. You must protect interior insulation from potential damage and decreased efficiency due to moisture. Eliminate all moisture problems before you close up the walls. Remember: If not protected against moisture, most building materials will not last.

WHAT TO USE: EXTERIOR

Installing the insulation on the exterior of the foundation wall increases demands on performance, structural integrity, and durability of the insulation. Unfortunately, many of the most popular insulating materials are useless as external insulators. Research on earth-sheltered dwellings provides us with guidelines for determining which products are best suited for application outside the foundation. Outside insulation should have:

- high compressive strength to withstand the weight of earth leaning against it
- high resistance to water in order to maintain R-value
- lifetime resistance to acidity found in some soils
- dimensional stability
- tongue-and-groove configuration to reduce heat leaks
- low cost, wide availability, and easy handling

In addition, most foundations have an above-ground area where the insulation must be protected from the ultraviolet rays of the sun. These rays break down foam insulations and can reduce them to dust unless protected.

Exterior insulation comes as rigid foam boards. You must choose between three types: Styrofoam, bead board (expanded polystyrene), and urethane. There is some confusion about the use of Styrofoam as a generic term, so be careful to understand the difference and purchase the correct product. Styrofoam is the brand name for an extruded polystyrene manufactured by the Dow Chemical Company. Styrofoam is blue in color, has a smooth surface, and can be bought with edges molded to fit together in a tongue-and-groove fashion. The white, pebbly, coffee-cup type of material is bead board. Bead board is often wrongly referred to as Styrofoam even by those in the trade. The third material, urethane, is most often tan, yellow, or green. It can be purchased with a reflective foil covering on both sides, which increases its performance as an insulator, protects the polyurethane from sunlight, and resists water damage. A fourth type, polyisocyanurate, sold under the trade names of High-R and Thermax, is similar to urethane except for a lower degree of flammability.

The long-term R-value of these foams is dependent on their ability to resist moisture. Absorption of moisture results in decreased R-value. Table 3.1 indicates the reduction in R-value due to the absorption of moisture after ten years in the ground.

Table 3.1

	Initial R-Value per Inch	Final[1] R-Value per Inch	Cost[2] per Board Ft.	Final R-Value per $
Extruded polystyrene Styrofoam	5.0	4.5	.28	16.1
Expanded polystyrene bead board	4.0	2.8	.12	23.3
Polyurethane	6.25	3.03	.45	6.7
Fiberglass		3.2 (for comparison only)		

This information based on study by Dechow and Epstein of the Dow Chemical Company. It has been updated with regard to prices by the authors.
1. Final R-value for 10 years exposed to ground.
2. Material cost per board foot based on 1981 Dodge Manual.

From Cornerstones Energy Audit, 1980

All things considered, extruded polystyrene—Styrofoam—stands out as our product of choice. It has high compressive strength, resistance to water and soil acids, a relatively high R-value per inch, dimensional stability, easy handling, wide availability, and tongue-and-groove edges. Expanded polystyrene does not have the overall strength of Styrofoam. It also does not come with a tongue-and-groove joining system, often breaks in handling, and absorbs water. Polyurethane, beyond being the most expensive of the rigid foams, also has absorption problems.

WHAT TO USE: INTERIOR

Interior insulation is not subject to the harsh environmental factors facing exterior insulation, but it has its own set of important guidelines to consider:
• high resistance to fire and/or protection from poisonous fumes during fire
• good dimensional and R-value stability
• high R-value per inch (especially if interior space is a factor)
• low cost, wide availability, and ease of installation and handling
There are several methods of insulating your foundation walls from the inside. The best method and material for you depends on the intended use of the area and your specific foundation system, especially the foundation material and condition of the walls.
Table 3.2 illustrates some of the more common situations and our recommendations.

HOW MUCH INSULATION?

As we have already stated, the earth around a foundation has a moderating effect. The degree to which the earth moderates heat loss depends on the amount of exposed wall and the relative conductivity of

Table 3.2

Basement Use	Special Considerations	Condition of Wall[1]	Procedure	Material
Living space	Basement being remodeled, new stud walls	Any	Install in stud spaces, finish	Fiberglass batts
Living space	Remodeling, limited space	Rough	2 × 2 framing insulation in bays	Rigid foam[2]
Living space[3]	No need to build new interior wall	Smooth	Adhere to wall, secure with gypsum[4]	Rigid foam
Crawl space or storage space	Appearance not a factor	Any	Hang down wall, secure at bottom, staple together	Fiberglass batts

1. Condition of wall—Smooth: block or poured concrete. Rough: stone or rubble.
2. Rigid insulation: bead board, Styrofoam, or polyurethane.
3. Basement use could be either Living or Storage for this method.
4. ½″ gypsum board is required over interior rigid insulation by fire code.

the soil. It may sound a bit odd to speak of the conductivity of soil, but heat loss below ground is primarily by conduction. Foundations pinned into ledge or in contact with moving groundwater are constantly carrying off heat and lose considerably more heat than a foundation set in dry, light soil. If you have either of these situations—ledge or moving water—you should attempt to form a thermal break between them and your heated space either by insulating on the inside in the case of the ledge, or by setting up proper drainage in the case of the groundwater (a professional job).

Illustration 3.1 demonstrates the effectiveness of insulating basement walls. It also shows that it makes good sense to use more insulation toward the top (exposed) portion of the wall. As you probe deeper into the earth below the frost line, the temperature rises toward a stable temperature that is the same as the local average annual temperature. (A rule of thumb is: Temperature average = 90°F minus your latitude.) Understanding this moderating ability of the earth helps you to decide the amount of insulation necessary for your climate. Generally, use half as much insulation as you would use for above-ground walls.

Home owners in the southernmost U.S. should *not* insulate their foundations, since homes there require more cooling than heating during the year. The "heat sink" effect of the earth, acting to wick heat from the structure, is an asset if you spend most of your energy bill to cool off. And that, by the way, brings up an interesting

Illustration 3.1

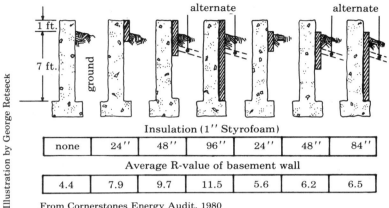

	Insulation (1″ Styrofoam)					
none	24″	48″	96″	24″	48″	84″

Average R-value of basement wall

4.4	7.9	9.7	11.5	5.6	6.2	6.5

From Cornerstones Energy Audit, 1980

point about the history of basements. Full basements are a traditional part of the northern home because builders have to dig deep, sometimes four or five feet, to plant a foundation below the frost line. (Soil, when it freezes, expands. A foundation placed above it would heave and crack like a frost heave on a road.) As long as the foundation has to go so deep, why not dig a couple of feet deeper and make a full-height underground room? Hence the full basement. But a deeper foundation isn't necessary in the Deep South, so a home foundation there traditionally has been a crawl space or a slab on grade (ground-level concrete pad).

INSULATING THE OUTSIDE

First, some heartening news. If you're insulating a full basement from the exterior, you do not have to excavate all the way down to the bottom of the foundation wall. In fact, you do not even have to excavate down to the frost line. You can "fool" the frost by insulating down and then away from the foundation. This interrupts the penetration of cold through the earth. You need only dig 12 to 18 inches deep and two to four feet out from the wall to install Styrofoam insulation (see Illus. 3.2).

Before digging, locate the utility lines servicing your home (gas, electric, telephone) by calling the utility companies. Most lines are either overhead or buried deeply out of danger, but it's worth the call.

Now that you're ready to insulate, here are the steps to follow:

Step 1. Determine how far down the outside of your foundation walls you want to insulate. Even running insulation just a few inches below the outside grade line will help reduce heat loss.

Step 2. If you plan to insulate the wall from the bottom of the siding to 12 inches below grade (ground level), run a 24-inch-wide board of Styrofoam horizontally away from the wall. Do this only if you

19

Illustration 3.2

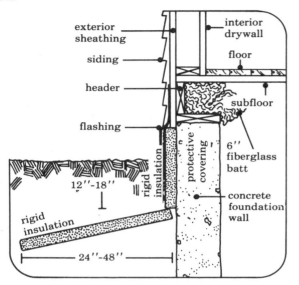

exterior sheathing

siding

header

flashing

interior drywall

floor

subfloor

rigid insulation

protective covering

12''-18''

rigid insulation

24''-48''

6'' fiberglass batt

concrete foundation wall

Illustration by George Retseck

do not plan to have any deep-rooted plants next to your foundation.

Step 3. Dig out the soil to the depth you have determined.

Step 4. Decide which protective covering you will use. If you have less than three feet of foam that will be exposed above grade after backfilling, a simple and inexpensive method for coating the exposed foam is to brush on two coats of liquid latex cement. For larger exposed areas, or areas where impact damage is likely, thicker reinforced cement coatings, stucco, or panels of cement asbestos board or exterior plywood are necessary. Your protective covering should extend four to six inches below the soil line.

Step 5. Apply a bead of caulking to the underside of the bottom edge of your house siding. Nail a prepainted or paintable wooden drip cap into the bottom of the siding, squeezing against the caulking you have applied. The caulking and drip cap will prevent water from going behind the Styrofoam.

Step 6. An alternative to Step 5 is to use painted metal upside-down J-channels instead of drip caps. J-channels can be purchased in various sizes at drywall supply houses. The size you need depends on the combined thickness of the foam board plus the protective covering you are using. Masonry fasteners applied about every four lineal feet will secure the J-channel firmly in place.

Step 7. Cut the insulation to fit around any surface projections such as windows or crawl space vents. This is done easily with any sharp knife. Just score and snap the insulation, or cut all the way through it.

Step 8. Apply Dow Insulation Mastic No. 11 or equivalent (read the cartridge's fine print to determine compatibility) to the insulation in ribbons, according to instructions on the mastic cartridge. If you're putting the Styrofoam over freshly applied waterproofing, wait three or four days for the waterproofing to cure. If you reside in termite territory, a liberal application of termite poison now between the Styrofoam and the foundation wall will keep the little critters away for the life of the foam.

Step 9. Do one board of Styrofoam at a time. When the mastic has been applied, place the insulation against the wall surface. Press the insulation uniformly against the wall, making sure the tongue-and-groove edges of the Styrofoam panels are snugly fitted together. Apply pressure over every square foot to assure a permanent bond. Where the insulation extends below grade, the backfill will hold it in place against the foundation wall. If desired, largehead masonry nails may be used instead of adhesive to secure foam to the foundation.

Step 10. Apply liquid latex cement to exposed area. A commercial mortar mix, such as Sakrete Mortar Mix, can be mixed to a brushable consistency in a pail, using a liquid latex, such as Acryl 60 (Standard Dry Wall Products, Miami, FL 33166, toll-free number 800-327-1570) or Laticrete 3701 (Laticrete International, Inc. Woodbridge, CT 06525, toll-free number 800-243-4788), or Camp Latex Silicone Liquid (Camp Company, Chicago, IL 60620, 312-779-4900). First, use a wire brush to thoroughly scratch the surface skin of the Styrofoam to be coated. Then mix up your latex cement, and apply one coat, using a four-inch texture paint brush such as Sears #30-12608. Allow this coat to get hard, then apply a second coat of freshly mixed latex cement. Do NOT use polyvinyl acetate latexes or powders, and do NOT apply latex cement coatings when the temperature will drop below 40°F in the next 24 hours. Do NOT add any water to your latex cement mix. Cracking of the latex cement coating over foam joints can be minimized by first caulking the joints or by embedding a thin fiberglass mesh in the latex cement applied over the joints. After a few years a fresh coat of latex cement may be desirable, to cover hairline cracks and accidental dents.

Premixed formulations for this application are available. Two such products are Akona Thermaseal (Akona Corporation, Maple Plain, MN 55359, 612-479-1907) or Insul/Crete (Insul/Crete Company, Inc., McFarland, WI 53558, 608-838-4541).

Step 11. Replace the excavated soil and you're done.

FURTHER CONSIDERATIONS

If a portion of your foundation wall is outside an unheated space, such as a garage, don't insulate the exterior of this area. The insulation should go on the inside, against the common wall between the basement and the garage. With living space above an attached garage, the ceiling of the garage should be insulated. Over an attached crawl space, the crawl space walls should be insulated, as specified in Table 3.2. Concrete or stone porches and paved driveways along the foundation walls often interrupt exterior insulation. In these situations the insulation must be installed on the inside. When insulating inside/outside always overlap the insulations at least two feet. Insulating a bulkhead door—a door leading from the basement to the outside—will be discussed in Chapter 4.

INSULATING THE INSIDE

If you're one of the lucky few who have no moisture problems and smooth foundation walls, you can insulate directly against the wall. If you plan to use rigid foam, follow these steps:

Step 1. Be sure the basement wall surface is structurally sound, clean, and dry. It must be free of dirt, grease, loose paint, and wallpaper.

Step 2. Remove baseboards, moldings, and window and door trim from all walls to be insulated.

Step 3. Nail horizontal wood nailers continuously along top and bottom edges of the walls and around untrimmed window and door openings. A ribbon of mastic can also be used behind the nailers to help hold them.

Step 4. Cut the insulation to fit around any surface projections, such as windows, nailers, electrical outlets, and conduits. This is accomplished easily with any sharp knife. Just score and snap the insulation, or cut all the way through it.

Step 5. Apply mastic to the insulation in ribbons, according to instructions on the mastic cartridge.

Step 6. Do one board of foam at a time. When the mastic has been applied, place the insulation (horizontally or vertically) against the wall surface. Press the insulation uniformly against the wall. Apply pressure over every square foot to assure a permanent bond.

Step 7. Do not insulate over water or drain pipes. During winter months, heat from the interior of your home may be necessary to keep the pipes from freezing and bursting. So butt the insulation up to the pipes. If possible, wedge some insulation behind the pipes. Bridging the pipes with ½-inch drywall is acceptable.

Step 8. When all the surfaces to be insulated have been covered with foam, you are ready to apply paneling or drywall. Apply mastic to the drywall in the same manner as you did the insulation. Adhere the drywall directly to the insulation. Apply pressure over every square foot of surface. Fasten the drywall to the wood nailers with drywall nails on six-inch centers. Put up all the drywall before going to the next step.

Step 9. Compound, tape, and sand the drywall joints according to the manufacturer's directions.

Step 10. Reinstall baseboards, moldings, and trim.

If your situation calls for fiberglass insulation, it can be installed in two ways: integrated into a stud wall for a finished basement, or hung from the floor joists or strapping in a crawl space or unfinished

Illustration 3.3

crawl space/basement

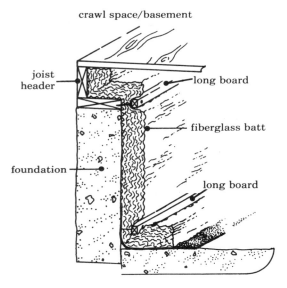

joist header

long board

fiberglass batt

foundation

long board

storage-type basement. The studding of the wall should accommodate the insulation — 2 × 4's, 24 inches o.c. (on center). Staple the insulation, (unfaced rolls or batts), into the bays between the studs and cover the warm side with a 6-mil polyethylene vapor barrier as you would do for the upper-level walls of your home. If you're hanging the insulation from the joist header, Illustration 3.3 shows you how. The insulation, foil-faced or kraft-backed in this instance, is tucked up into the header area and secured with a long board and nails. Place it with the backing toward you, not toward the wall. Allow the insulation to drape down the wall and secure it at the bottom with another long board to prevent air from circulating behind the insulation. The batts of insulation should be joined together by unfolding the flanges of the backing and stapling them together.

If you have a moisture problem you can't remedy and you can not insulate on the outside, you can protect the fiberglass with polyethylene. If the moisture leak is along the floor, insulate down to within four to six inches of the floor. (Remember, your foundation walls lose very little heat at the lower depths.) The insulation should be protected from any moisture migrating up either the wall or the framing lumber to which it is attached. In the worst of situations it is necessary to wrap both sides of the insulation to keep it dry and effective, so in this situation polyethylene is necessary against the foundation wall as well as on the warm side of the insulation. If moisture is a recurring problem in your basement, use only pressure-treated framing lumber.

Chapter 4

Insulating Your Walls and Floors

THE TIGHTER HOUSE

Insulating exterior walls is a large and difficult task. The area to be insulated is usually two to four times that of the attic, and the insulation options usually are neither clear nor simple. The first task is choosing an insulation material and technique. A careful inspection and evaluation of the condition of your home is critical to deciding which of the following options is best for you. Table 2.1 tells you more about these products. Here is what's available:

* loose-fill insulation blown in from outside
* loose-fill insulation blown in from inside
* fiberglass batts or blankets
* loose-pour vermiculite or perlite
* blown-in foam (urea-formaldehyde)
* exterior rigid foam boards
* interior rigid foam boards

WHAT'S YOUR SITUATION?

Interior good, exterior good. Your home is in good repair inside and out, and you do not wish to alter its appearance. Loose-fill cellulose or fiberglass blown in from the outside by an insulation contractor is most likely the cheapest, least disruptive method of insulating. Selected shingles or clapboards are removed and holes drilled through the sheathing in order to blow the fill into the wall cavity. Filling the 3½- to 4-inch space results in a total wall R-value of 11 to 15, a cost-effective way to preserve the existing appearance of your home.

A possible alternative is blown-in urea-formaldehyde foam. Proper installation is extremely important with this material. Improper installation often leads to excessive shrinkage (resulting in heat bypasses within the stud spaces and low effective R-value), paint peeling on inside or outside walls, and an objectionable level of formaldehyde odor. The Consumer Product Safety Commission is investigating the formaldehyde foam issue. By the time you read this book, it may have been banned or subject to stringent application guidelines.

Interior good, exterior needs work. If the exterior of your home needs only minor repairs or just a good paint job, your best option still is blown-in loose-fill. It is easier to justify removing clapboards and drilling holes in the side of your house if you already plan to repair or paint. Combining the two jobs saves you time and money.

If the exterior siding of your home is in poor condition and is no longer serving its primary function of keeping out wind and rain, you may consider new siding. In that case, ¾- to 1½-inch exterior rigid foam insulation (polystyrene, urethane, or isocyanurate) can be applied right over the existing siding and the new siding over that. Of course, any moisture problem that may have led to a premature deterioration of the existing siding in the first place should be eliminated prior to insulating. Vinyl and aluminum siding contractors are equipped to do this job. If the wall cavity is to remain empty or if a vapor barrier is already in place, a new vapor barrier is not necessary. Spaces between the old siding and the foam serve to ventilate the moisture out to the corners of the building where it can escape. Fortunately, in all but the very coldest climates (8,000 degree days) the dew point is most likely to occur within the foam, where there is no moisture to condense.

Interior needs repair, exterior good. The old plaster, wallboard, or paneling has deteriorated or is just not to your liking. Your options now switch to the inside and can be approached in two ways. One approach is. to completely gut (tear out) the interior surfaces of the outside walls. This will leave you with exposed wall cavities waiting to be insulated and refinished, a better understanding of how your home was built, and a nose full of plaster-dust and dirt. The existing studs should be extended to accept 5½ to 6 inches of fiberglass insulation, a vapor barrier, and a new interior surface.

The second approach is to drill holes through the interior sides of the walls and blow in loose-fill insulation. A polyethylene vapor barrier can then be stapled over the existing wall surface and a new finish of gypsum drywall installed. The second method has two advantages: it eliminates the time, labor, and mess of gutting; and loose-fill fills unevenly spaced stud cavities more easily than batt. With a very careful drywall job, window and door trim may not have to be adjusted, but the trim will lose some of its relief. Of course, electrical outlets will have to be extended out from their position in the old walls so that they fit flush against the new drywall.

Interior new cottage style, exterior good. With new construction, fiberglass batt is the easiest and most cost-effective material for insulating. In a cottage, post and beam, or cabin-style structure with uncovered wall cavities, there are two options: interior fiberglass and exterior rigid foam. If the interior is attractive and you wish to preserve the rustic appearance, insulate the exterior, using foam. If not, insulate the inside using fiberglass batts, and cover them with new drywall or paneling.

Masonry buildings are the most difficult residential buildings to insulate. Exterior insulation is normally impractical for appearance reasons, and if there is a cavity in the middle of the masonry walls, it is most likely very narrow and susceptible to moisture penetration during driving rains. Drilling into such cavities is difficult and expensive, and restoration of the walls to their original appearance after the work has been done is chancy as well as time consuming. In rare cases the tops of the wall cavities are accessible from the attic. Loose-pour vermiculite (expanded mica) or perlite (glass beads) can be poured into the cavity with the hope that it goes all the way to the bottom. Otherwise, filling the space within the walls is definitely a job for professionals. This is an expensive job with a long payback period.

If you are a bit more flexible about the appearance of your home, two options become available, one on the inside and one on the outside. Rigid foam insulation can be applied to the interior plaster or block wall with a mastic adhesive and then covered with ½-inch gypsum drywall exactly as described in Chapter 3 for basement walls. This method provides insulation and preserves the exterior brick or stone finish. Unfortunately, the thermal mass of the masonry (very influential in moderating temperature swings and holding heat) is lost. The trim and wiring also have to be redone.

Insulating the exterior of masonry walls with rigid foam is becoming more common. There are several patented and franchized systems, such as Drivit and Thermalbar, for applying a masonry-type finish over expanded polystyrene (bead board). Do-it-yourselfers and siding contractors can install any type of siding over any type of rigid foam by one of the methods described below.

HOW TO INSULATE WALLS

A strong word of caution: Before insulating walls, have your electrical system inspected by a qualified professional. The Consumer Product Safety Commission estimates that more than 2,000 household fires occur each year because thermal insulation is ignited by faulty wiring. Heat build-up within wires surrounded by insulation is a potential problem in the walls and attics of all homes. Homes built before 1940 are of particular concern because the rubber-based insulation covering the wires may have deteriorated to the point where handling or movement could expose the live wires. A presently stable situation can be aggravated by the installation of insulation. The cost of a typical inspection is between $25 and $50 per home, and the cost of repairs can range from zero to $1,000 depending on the situation and the size of the home.

Of the available options for insulating your walls, most are best installed by the professional contractor. If you decide to call in an insulation contractor, be sure to agree to terms, in the form of a written contract, before the job is begun. A wise addition to a wall-insulating contract would be a provision for a thermographic survey (infrared photography) of the finished job. The contract should provide for correction at no cost of all gaps or missed cavities revealed by such a survey within one year after the job is completed. It is very important for both you and your contractor to state the limits of the work, especially with regard to thermal quality and the degree or restoration and cleanup.

Fiberglass Batts or Blankets. Fiberglass batts or blankets (see Illus.

Illustration 4.1
Batt or Blanket Wall Insulation

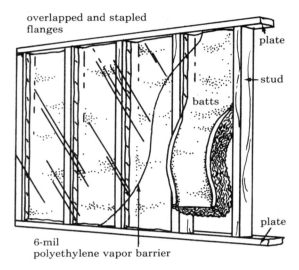

overlapped and stapled flanges

plate

stud

batts

plate

6-mil polyethylene vapor barrier

4.1) are the easiest and most common do-it-yourself materials for insulating walls. With exposed interior wall cavities your task is simple: fill these cavities completely with insulation. Your work area should be clean and open to insure ease of installation and safety. Loose-fitting work clothes closed off at the neck, wrist, and ankle, gloves, goggles, and a particle mask are the fashions of the day. For tools you'll need a tape measure, sharp knife, straightedge (a 1 × 6-inch board will do), hard cutting surface, hammer, and a heavy-duty staple gun.

While the studs are accessible, extend them to 5½ to 6 inches (most walls are of 2 × 4-inch construction). Extending the studs is accomplished simply by nailing 2 × 2-inch "nailers" or "scabs" atop each stud. Nailers are available at any lumberyard. The new 6-inch depth allows you to use more insulation, increasing the R-value from R-13 (3½-inch) to R-20 (6-inch wall). In addition to building out the studs, the casings around windows and doors should also be extended and the electrical outlets and switches built out. The extra labor and materials will be returned in energy savings and increased comfort.

If the studs are spaced evenly at 16 or 24 inches from center to center, the job is simple. Unfortunately, the stud spacing in many older homes is not standardized. With irregular spacing it is best to buy batts larger than the space and cut the batt ½ inch wider than the space so the fiberglass fits snugly. After unrolling and laying out the insulation, cut the length also about ½ inch longer than the space. If you buy foil or kraft-backed insulation, install it with the backing toward you. There are flanges on both edges for stapling. Fold these flanges flat over the

27

studs on either side and staple from top to bottom, being careful to keep the flanges flat and the insulation within the cavity. To avoid all-important heat bypasses, all gaps or spaces should be carefully stuffed with pieces of fiberglass. (You'll have lots of scrap!) Next, staple a continuous 4- or 6-mil polyethylene vapor barrier across the face of the studs, starting in an upper corner. Keep the polyethylene sheet taut. Now cut the drywall to shape, leaving room for electrical boxes, windows, and other penetrations. Install the drywall, *then* cut out the polyethylene where it covers penetrations. This will insure a tight seal. The final phase is painting or wallpapering and installation of foam rubber gaskets and cover plates on the electrical outlets.

Loose-Fill Insulation Blown in from Inside. We don't recommend that you attempt this job yourself unless you are very experienced or can obtain some experienced help. Location of wall blocking can be tricky, and proper installation density is critical to avoid long-term settling. However, equipment can be rented at tool rental stores. As with any rental equipment, make sure that you've done everything you can do before the equipment arrives. Drill holes in each cavity, making them at least as big as the installation hose you'll be using. Make the holes with a keyhole saw or a saw bit on a drill. Then build out all electrical boxes. The holes should be at the top of each cavity—the top of the wall and just below the fire stops, if you have them. A plumb bob or a weighted object on a string inserted into the cavity will help you find fire stops or other obstructions. The locations of studs can be determined by inserting a short length of rubber hose sideways into each hole after drilling.

In renting the blower, remember that you can often get a "free" day if you rent over the weekend, which allows you more time and care. When you pick up the blower, ask the salesman as many questions as you can think of. If you can get a free demonstration, especially on your home, do so. Once you get the unit home, double-check to make sure the equipment and the loose-fill insulation are compatible. For cavities up to eight feet high, install 3½ pounds per cubic foot; for cavities over eight feet high, use 4½ pounds per cubic foot. Beyond these directions there is little more to offer; you will learn by doing. The operation requires a team of three: one to work the hose, one to load the bags of fill into the blower and one for general assistance and moral support. It takes a while to get used to the blower, but by the end of the job you'll feel like an old pro.

After filling the wall cavity with insulation, staple a continuous sheet of 4- or 6-mil polyethylene over the old surface, nail up ⅜-inch drywall, and finish the wall with paint and wallpaper. (Putting up new drywall will be easier than repairing all the holes in your wall and will give you the chance to install a good vapor barrier.) Measuring and cutting the drywall very carefully around windows and doors minimizes the spackling and finish work later.

Interior Rigid Foam Boards. Installing rigid foam insulation on the interior surface of your walls can be done either by a professional contractor or by a skilled home handyman. The skill is not necessary for installing the insulation, but rather for performing the trim work around the windows and doors. Trim should be built out with nailers

28

before the insulation job begins. Nailers of the same thickness as the foam should also be installed at the very top of the wall and at the very bottom. The electrical boxes must also be brought out to sit flush with the new finished surface (1½ to 2 inches of insulation plus ½ inch of drywall).

The rigid insulation can be applied directly over the existing wall surface with an adhesive recommended by the manufacturer of the foam. It is wise to purchase the foam and the adhesive together from a reputable insulation dealer. The insulation boards can be applied in a single or a double layer. If two layers are used, the second layer should cover the joints of the first. Rigid insulation is cut with a handsaw or repeated passes with a sharp, heavy-duty utility knife. The same adhesive used to attach the insulation to the existing wall can be used to fix the drywall to the insulation. Nail or screw the drywall to the wood nailers around windows and doors, and to those at the top and bottom of the wall.

Exterior Rigid Foam Boards. Rigid foam insulation can be applied over wood or masonry exterior wall surfaces and covered with either stucco or siding. The recommended method consists of fastening metal Z-channels (as made by U.S. Gypsum or equivalent and sold at drywall or siding supply houses) in vertical rows spaced 24 inches on center using ¾-inch-long air gun-driven nails. Install a horizontal wood starter strip at the bottom of the Z-channels. Next, slip into the channel 24-inch-wide foam boards of the same thickness as the Z-channel. Finally, install the siding.

If the siding is vinyl or aluminum, use self-tapping metal screws. If the surface finish is to be stucco, first install 3.4-pound galvanized, ⅜-inch self-furring expanded metal lath, and then apply stucco in a total thickness of ¾ inch to 1 inch.

Loose-Pour Vermiculite or Perlite. If masonry walls are hollow and open to the attic space, you can pour vermiculite or perlite into the cavity. (Before you begin, lower a weighted object on a string to determine if the cavity extends all the way to the foundation. If it does not, half a job is not worth the trouble and another insulation option should be explored.) Both products come in large, light plastic bags with charts and tables for determining the quantity needed to complete the job. Open the bags one at a time or you'll have pellets all over your attic. In most situations it is worth your effort to construct a makeshift trough or funnel to direct the material into the cavity.

INSTALLING FLOOR INSULATION

If the basement below the first floor is not utilized as a living space, you may wish to insulate the floor beneath your feet rather than the basement walls. Whether to insulate the basement walls or first floor is determined by the anticipated use of the basement and the potential for freezing pipes. Generally, if the hot water and heating systems are in the basement, it is best to insulate the basement walls. If this equipment is not in the basement and you do not use it as a living space, insulate under the first floor. This reduces the total volume of heated air within the house and requires fewer square feet of insulation to do the job. Gravity dictates the type of material to use in under-the-floor insulation. Loose-fill is out, which leaves only rigid foam and fiberglass batt. In most situations, because of its flexibility and low cost, unfaced fiberglass is the material of choice. Insulating under a floor is not an easy task. In most cases you'll run into cross bracing, wiring, plumbing, flooring nails, and the uncomfortable situation of working over your head, with dust, dirt, and fiberglass falling onto your face.

On the brighter side, floor joists are more apt to be evenly spaced, and the depth of the joists is usually large enough to hold plenty of insulation without alteration. The thickness of the insulation in the floor can be less than in the attic because of the tempering effects of the earth around the foundation. A properly designed and constructed basement with a limited amount of exposure (12 to 18 inches above grade) needs only half as much insulation as the roof of the same building to save the same amount of heat. In all but the coldest of climates, 6 inches of fiberglass in either the floor or the basement wall is adequate.

Insulating between evenly spaced joists, 16 or 24 inches on center, is simple. Due to the earth-tempered temperature of the basement, a vapor barrier is not required. Buy unfaced batts of fiberglass of thickness equal to or greater than the depth of the floor joists. Hold them in place against the subfloor with lengths of thin wire or nylon fishing line stapled to the joists in a zigzag pattern. It is easy to see that some assistance is welcomed for this task.

Uneven floor joists require customizing the batts. Cut the batts ½ inch wider than the space between joists and hold them firmly but gently against the subfloor with wood strapping, nylon line, galvanized wire,

Illustration 4.2
Fiberglass Batts Secured under Flooring

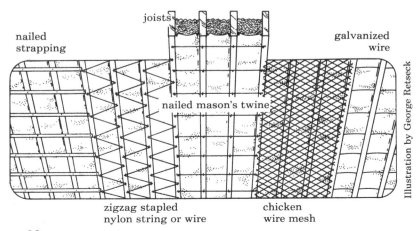

joists

nailed strapping

galvanized wire

nailed mason's twine

Illustration by George Retseck

zigzag stapled nylon string or wire

chicken wire mesh

or chicken fencing. To avoid bypass heat losses, the batts of insulation should fit snugly against both floor and joists. The flexibility of the fiberglass helps hold it in place by friction until you can secure it with one of the methods shown in Illustration 4.2.

Wood strapping spaced 16 inches on center can hold the fiberglass in place if the batt is at least as thick as the joist depth. Wood strapping also makes the installation of a finished ceiling simple if you decide to finish the basement at a later date. Nylon string or thin wire can be stapled in a zigzag pattern from one joist to the next to hold the insulation in place. Mason's twine can also be used. Chicken-wire mesh provides the most security, but it's also the most expensive and difficult to work with. (Buy 2-inch mesh: it costs one-half as much as 1-inch mesh.) Finally, 15- and 23-inch lengths of galvanized wire with sharpened ends are available from insulation retailers and wholesalers. Either cut these wires to fit the unevenly spaced joists or install them at an angle.

The choices of the best material and installation technique are very important. If you do not feel comfortable with your decision, do not hesitate to call a number of professionals. Insulation contractors give free advice, and calling at least three with good reputations may indicate by consensus the proper selection.

▄▄Chapter 5▬▬▬▬▬▬▬▬▬▬▬▬▬▬▬▬▬

 **Insulating the Attic**

| THE TIGHTER HOUSE |

My mother always told me to put a hat on my head as I left the house on a cold winter's day. I didn't always listen, and I came down with my share of colds. What did my mother know that I was too stubborn to realize? The top of your head, like the top of your house, is the greatest area of heat loss. Heat rises to these areas that are directly exposed to the wind and cold.

The attic is often the easiest and cheapest area to insulate, and it's almost always a job you can do yourself. The payback period for your money spent insulating an uninsulated attic can be as little as two years.

Your aim is simple: to retard the flow of heat from the heated space below to the unheated area above. If the attic is used only for storage, it need not be heated, so the insulation should be placed on the ceiling directly above the heated space, not between the attic rafters. With insulation here, and with proper ventilation, your attic temperature should be close to the outdoor temperature, both winter and summer.

Care should be taken from the beginning to insure safety and performance. Proper moisture control practices such as vapor barriers and ventilation are especially important in the attic. A conscious effort must be made to install the insulation so it does what it's supposed to do. A few preinstallation precautions will increase the insulation's overall performance.

BYPASS LOSSES

Many people who have insulated attics have expected tremendous savings based on predictions made from heat loss calculations. Unfortunately, the predictions didn't come true. Recent studies conducted by the Center for Energy and Environmental Studies at Princeton University indicate why most insulated homes are not conserving as much heat as they should.

The Princeton studies show that heat losses through attics are often three to five times greater than calculated predictions. The discrepancy is not due to false advertising from insulation manufacturers regarding the thermal capabilities of their products. Rather, the reduction in performance is due to careless installation practices and oversights that leave paths open for warm air to escape. The heat is not traveling through the insulation, but is bypassing it. These areas of hidden heat loss are similar to the infiltration losses around unweatherstripped windows and doors.

Many of these bypass areas can be identified and eliminated. A careful examination of the attic can reveal potential heat leaks around plumbing vents, electrical wiring, recessed lighting fixtures, furnace flues, masonry or metal chimneys, interior walls, and at the intersection of the attic floor and the exterior walls. Large openings can be stuffed with a fireproof material such as unfaced fiberglass insulation. (Do not use faced fiberglass because the facing is flammable.) Smaller openings can be sealed with caulking, duct tape, or polyethylene.

Other common but not so obvious bypass areas occur above stairways and dropped ceilings. These leaks are not always visible from the attic but still offer warm air pathways to the attic. Finally, the most commonly overlooked area is the trap door or scuttle hole into the attic. This door should be weatherstripped as if it were an exterior door and insulated equally with the rest of the floor.

Your attic probably has its own unique bypass paths. Take the time to find and seal them. Obviously, the job is easier and more effective when done before the attic is insulated. Fixing attic bypasses usually costs no more than $15 but may return as much as 15 percent of your total space-heating cost every year. Fifteen percent may seem a bit large for such a small investment, but as the major heat losses are eliminated the smaller leaks become increasingly more significant.

We will look at attic bypass losses in more depth in Chapter 9.

WHICH INSULATION TO USE

The amount of insulation to add to your attic depends on three things:
• your climate
• the amount of insulation already present
• the type of insulation you choose

Illustration 5.1 is a map that indicates the recommended R-values for different climatic zones. Find your zone and use the accompanying table to determine the proper amount of insulation for your attic.

To find out if you already have insulation beneath a floor, you must either pull up a floorboard or drill a small hole through the existing flooring. What you are likely to find is rock wool. Although rock wool looks dirty, leave it; it is excellent insulation and is no doubt still performing its job.

Table 2.1 compared the most commonly used residential insulation products. For attics, the low-cost blanket, batt, and loose-fill products are best. Batts and blankets of fiberglass and rock wool are manufactured in standard widths to accommodate modern standardized joist and

Illustration 5.1
Heating Zones and Recommended R-Values

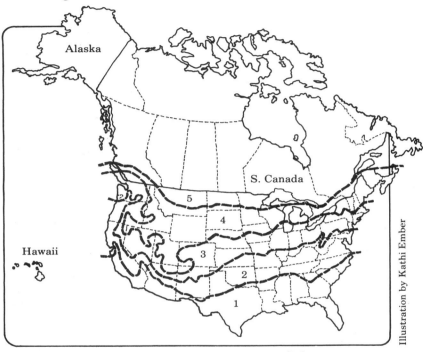

Heating Zone	Attic Floors	Exterior Walls	Ceilings over Unheated Crawl Space or Basement
1	R-30	R-value of full wall	R-11
2	R-30	insulation, which is	R-13
3	R-33	3½'' thick, will depend	R-19
4	R-38	on material used.	R-22
5	R-49	Range is R-11 to R-13.	R-22

rafter spacing. If your joists or rafters are spaced either 16 or 24 inches from center to center, the batts and blankets will be easy to install. If the joists are irregularly spaced, a loose-fill insulation is recommended.

Table 5.1 compares common batt/blanket and loose-fill types of insulation for their relative R-values per inch. From this table you can see that there are space limitation problems with some of the insulating materials. If you are trying to achieve R-38 with loose-fill fiberglass, for instance, the depth of insulation must be approximately a foot and a half. The joists in most attic floors, however, are normally only 2 × 6, 2 × 8, or 2 × 10. Remember, if you plan to have flooring over the

34

Table 5.1

R-values per Inch	BATTS/BLANKETS		LOOSE-FILL		
	Fiberglass	Rock Wool	Fiberglass	Rock Wool	Cellulose
R-11	3½''	3''	5''	4''	3''
R-19	6-6½''	5¼''	8-9''	6-7''	5''
R-22	6½''	6''	10''	7-8''	6''
R-30	9½-10½''	9''	13-14''	10-11''	8''
R-38	12-13''	10½''	17-18''	13-14''	10-11''

insulation, the insulation must attain the desired R-value within the depth of the joist.

Most people who live in Zones 1 and 2 (see Illus. 5.1) can insulate between their attic joists with fiberglass batts, rock wool or loose-fill cellulose, and achieve the recommended R-value without insulating above the tops of the joists. But those who live in Zones 3, 4, or 5 must either find another place to stow their Christmas tree decorations, or build a small platform atop the new level of insulation.

The cost of blanket and batt insulation per unit of R-value usually is slightly lower than loose-fill with standardized framing, and the installation is easy and quick. Most home owners prefer the batt (cut in four- and eight-foot sections) over the blanket (a continuous roll) because of its ease in handling. With irregularly spaced joists, cellulose loose-fill is the most cost-effective. Cellulose can also be blown in under a finished floor you don't want to remove. In simple situations, the home owner can rent a blowing machine from a tool rental outlet and do it himself.

PREINSTALLATION

Before you begin to install the insulation, a few steps should be taken to insure safety and success. Inspect the roof and the attic floor for signs of moisture, such as warping, stains, and visible fungus growth. Don't ignore these danger signals! Eliminate the source of these problems— most likely a leaky roof—and correct the unfavorable conditions before you insulate. Electrical wiring should be replaced if its insulation is old and frayed. Measure the area to be insulated so you can order the insulation all at once. (Shopping around by telephone for insulation is likely to save you a few dollars—as it does with all building products.) While shopping, inquire about free delivery. Most building supply outfits have free delivery within a 20-mile radius if your order is substantial.

A well-prepared work area makes your task much easier and far more enjoyable. If the floorboards are to be taken up, they should be removed from the area if possible; if not, they should be stacked neatly to one side. Adequate lighting should be supplied. Temporary flooring (movable boards) should be used so you don't end up falling through the ceiling into the living room. If you've decided to insulate with fiberglass you will need the following tools: a heavy-duty staple gun (a cheap one will only frustrate you), a pair of heavy-duty shears or a sharp knife, and a roll of duct tape to seal leaks you may encounter. If you're installing cellulose or any of the other loose-fill types you'll also need a garden rake or a wide board to spread the insulation. Wear clothing that fits tightly at the neck, wrist, and ankle but is loose otherwise, to provide protection but at the same time allow freedom of movement. A mask and gloves are recommended when using fiberglass. A hard hat or helmet are in order if roofing nails project through the roofing boards overhead.

Electrical components—wiring, recessed lighting, junction boxes— deserve careful consideration. The guaranteed heat bypass created by the National Electrical Code's three-inch clearance requirement for recessed lighting fixtures favors the elimination of all recessed lighting. If these lights are not removed, install permanent blocking of high-temperature plastic, wood, or light sheet metal around the fixture at least as high as the finished height of the insulation. Be sure this blocking does not inhibit future maintenance or servicing. Electrical cables should never be sandwiched between layers of insulation or fully surrounded by loose-fill insulation because overheating and fire can result if a circuit breaker fails to operate or the breaker is overfused. Therefore, make sure the cable is either under or over the insulation, but never between.

HOW TO INSULATE

The Unfinished Attic. If your attic space is not to be used as a living space, the insulation should go into the floor. If there is no flooring present in your attic you are quite fortunate. All that need be done is to install the insulation between the joists. If there is flooring in place, you may either rip up the flooring and insulate, blow in cellulose insulation under the in-place floor boards, or install insulation on top of the floor.

With standardly spaced attic joists you'll most likely choose to insulate with fiberglass. Unfaced batts are greatly preferred because they are more flexible than faced batts and fit more snugly into the joist cavities. Also, using faced batts poses the danger of trapping moisture within your insulation. The insulation should be unrolled as you need it, cut to length, and pushed into place between the joists, starting at the eaves (but not covering soffit vents) and working toward the center. Insulation should always fit snugly but should not be compressed. It should maintain its full thickness to insure maximum performance, because its insulative quality depends upon the amount of air it traps.

Cutting the fiberglass is simple. Roll it out on a flat surface; lay a board on it to serve as a straightedge; kneel on the board to compress the fiberglass; then cut it with a sharp knife. If you are unable to locate unfaced fiberglass to add to existing insulation, simply remove the facing.

Even with standardized joist spacing there are a few minor obstacles.

36

There is no guarantee that all the spaces will be 16 or 24 inches on center, especially around the chimney, the trap door, and the end walls. For spaces wider than the insulation, the full-sized batt should be moved to one side and the remaining space filled with either a cut strip of another batt, or with loose-fill insulation. If the insulation is too wide, it should be trimmed to fit, but be sure its final width is ½ inch wider than the space into which it is to be placed. The cross bracing found in many joist systems can also slow you down. Here, the insulation should be split down the center about 12 to 18 inches on either side of the braces. The batt should have a staggered cut where it meets the brace and should be installed in an over/under pattern. An even better solution is filling the braced area with loose-fill insulation.

Loose-fill insulation also is installed from the eaves in. A garden rake is the best tool for pushing the material into the eaves and for spreading the fill evenly between the joists. If you're renting a blower to help you spread loose-fill, be sure to operate the equipment in accordance with the manufacturer's instructions. The lowest pressure recommended is usually best for the do-it-yourselfer. If loose-fill is to be installed above the tops of the floor joists, leveling strings can help guide you in making an even distribution.

Be certain that the insulation, whether loose-fill or batt, does not block the soffit vents (see Illus. 5.2). One preventive technique is nailing 1 × 8 blocking along the inside edge of the eaves, as shown. If you're installing loose-fill, this blocking will also keep the small gusts that occasionally enter the soffit vents from playing havoc with your tidy distribution.

The door to the attic must not be overlooked. It should be weatherstripped and insulated. The attic side of the door can be insulated with a batt of fiberglass. A bulkhead-style box can be built

Illustration 5.2
Keeping Soffit Vents Clear

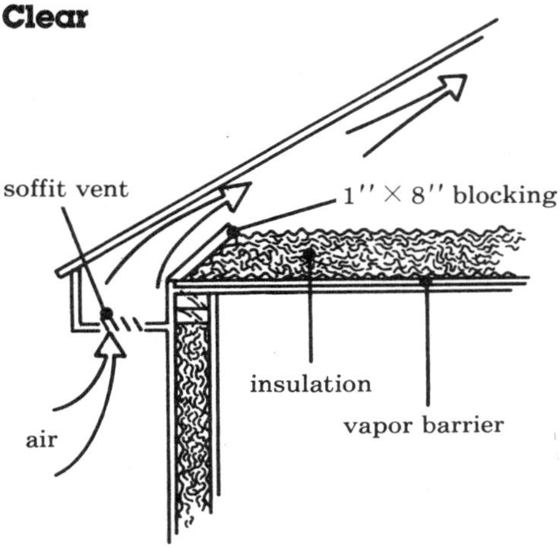

Illustration by George Retseck

soffit vent

1″ × 8″ blocking

insulation

vapor barrier

air

around this opening out of 2 × 4's and plywood. The top and all the sides must be insulated. If you have a whole-house fan, a well-insulated portable box can be built to fit over the fan during the winter and removed in summer. Also, a piece of fiberglass batt can be sandwiched between the shutter and the fan during the winter.

The Finished Attic. A finished attic is more difficult to insulate. Illustration 5.3 shows the insulation areas in a typical finished attic. The insulation is installed between the conditioned (living) spaces and the unconditioned spaces. Getting to the cavities between these spaces can present some problems. It may be necessary to cut one or two small openings in the knee-wall to gain access.

Illustration 5.3
The Finished Attic

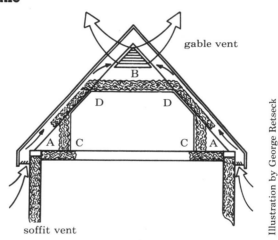

gable vent

soffit vent

Illustration by George Retseck

The outer attic floor and the attic ceiling (A and B in Illustration 5.3), if accessible, can be insulated as if they were an unfinished attic floor. If inaccessible, loose-fill can be blown into these areas by a professional. The knee-wall (C) can be insulated with fiberglass batts placed between the exposed studs of the wall. Unless the attic area is under construction and the end walls are exposed, the best material for the end walls is cellulose blown in from the outside. Again, this is a job for the insulation contractor.

The area along the rafter (D) presents the most difficult problem. It is hard to get to and offers a potential moisture problem if air is not allowed to pass over the cold side of the insulation. A 1-inch air space above the insulation should be provided for ventilation so that the roofing boards are not damaged by dry rot or overheating in the summer. In rare cases it is possible to shove fiberglass batts down the rafter spaces and maintain such an air space above the insulation. Otherwise, the proper solution is removal of the sloping ceiling and

38

installation of fiberglass, or else installation of rigid foam insulation on the slope, covered with ½-inch gypsum drywall.

Difficult Roofs. Your roof may not be as simple or as accessible as the ones mentioned above. Cathedral ceilings (mentioned below), flat roofs, or the attic spaces that are just too cramped to work in call for creative, difficult, and often more expensive methods of insulation. Insulation contractors, however, face these situations often and have clever solutions for most problems, so consulting a contractor is a wise first step.

The cathedral ceiling is a very common but very difficult insulation task. The best method of insulating depends on the style of roof construction and your aesthetic preferences. If your ceiling has exposed rafters spaced 16 or 24 inches on center and you don't mind covering them up, you're in luck. Install batts of fiberglass between these rafters (being sure to leave 1 to 2 inches above the insulation for air circulation), a 6-mil polyethylene vapor barrier below the insulation, and a finished ceiling of gypsum drywall or boarding. A soffit and ridge vent system can provide the necessary ventilation.

A cathedral ceiling with widely spaced rafters can be insulated on the inside with rigid board insulation, then covered with ½-inch drywall. If a vapor barrier is placed between the insulation and the drywall, ventilation above the foam is not required.

If your cathedral ceiling has purlins (horizontal framing members between the rafters), or if the interior roofing is too beautiful to cover, you must insulate from the outside. You'll have to remove the roofing material (probably asphalt shingles) and apply rigid insulating boards directly over the roofing boards. A lifetime roof can be achieved by first nailing a 3½-inch (2 × 4-inch) wood strip around the perimeter and at the peaks and valleys of the roof. The rafter locations should be marked on the wood strips for the nailing to be done later. Install a continuous layer of 6-mil polyethylene, followed by a 3½-inch thickness of rigid insulating board to fill the area between the wood nailers. Cover the insulation with plywood, ½-inch CDX, nailed to the wood perimeter strips and through the foam into the rafters below. Use 4½-inch-long galvanized box nails. Finally, install new roofing material over the plywood. Obviously, this is a job preferably saved for the day you need a new roof.

 **Doors:
When You
Shouldn't Bother**

| THE TIGHTER HOUSE |

Little thought is given to the heat loss of doors except with regard to infiltration and weatherstripping. For one thing, doors account for only 2 percent of the total wall area of most homes, and for another, the R-value of a normal wood exterior door is about the same as that of a triple-glazed window. Also, the costs of most commercial solutions designed as retrofits are so high that energy savings over the life of the product would perhaps never make up for the added cost. Incorporating special doors or insulating features into *new* construction, however, is not as expensive as retrofitting, and so the payback on such investments is rapid. Consider the following if you are building a new home or if your retrofit definitely involves trashing your present door. If the present door is still serviceable, it is wiser to put money into cushion-bronze weatherstrip instead (see Chapter 8).

Thermal Doors. The standard exterior door of today is a plastic or steel shell filled with urethane insulation. R-values range from 6 to 14 versus values of 2 to 3 for the old-fashioned six-panel or solid-core wood door so common in the past. Nonwood doors are not subject to warping or seasonal swelling and shrinking that used to make weatherstripping so difficult, and with the increasing price of high-quality wood, they are actually *less* expensive as well.

Don't rush out and buy one just to replace your wood door, though. You'll find you can't plane down a vinyl or steel door to fit your present opening, and chances are you'll have to buy a prehung unit that will replace your door frame at the same time. This makes it unlikely that you'll recover your investment with energy savings.

Storm Doors. Let's face it, a storm door is really an aluminum version of what we used to call the screen door. Although almost everyone today has storm doors, National Bureau of Standards tests have shown that they are effective only to the extent that they can be better weather-sealed against infiltration than the normal entrance door. The reduction in infiltration when a storm door is installed over a properly weatherstripped entrance door is nil. In addition, storm doors raise the total R-value of the entranceway a mere single unit—usually from R-2.5 to R-3.5—the

40

same amount as any single layer of added glazing. When you add to this the expensive standards to which a storm door must be built in order to withstand the inevitable abuse it receives, this popular and highly touted addition to a home's heat-saving equipment seems a poor investment indeed.

Air Locks. The advantages of air-lock entrances depend largely on the amount of exposure to wind the house receives, as well as the overall air-leakage characteristics of the house. An otherwise tight house nestled in the woods will not benefit significantly from an air lock. On the other hand, a sieve on top of a wind-swept hill is untenable without one. Unfortunately, air locks are best incorporated at the blueprint stage. Otherwise they look funny, occupy valuable space, and interrupt the intended traffic flow.

Rather than build one onto an existing house, a better solution is to caulk shut a windward door and start using a door off the garage or glassed-in porch exclusively in the winter.

SOME UNCOMMON SOLUTIONS

Even though prospects seem bleak for improving our front doors, there may be other doors you haven't even considered. Energy improvements to these doors are probably right up there with attic insulation in cost-effectiveness.

Garage Door. If your garage is heated or even semiheated, chances are the greatest heat loss is through that monstrous garage door. First you should weatherstrip it, using cushion-vinyl weatherstripping on the sides and a door sweep especially made for garage door bottoms (see

Illustration 6.1
Insulated Garage Door (Cross Section)

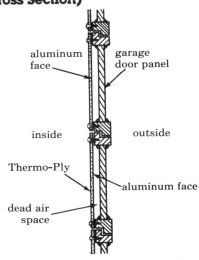

aluminum face

garage door panel

inside

outside

Thermo-Ply

aluminum face

dead air space

DOORS

Chapter 8). Then the following incredibly simple operation increases the R-value of the normal overhead panel-type door from 1.3 to 4.0 (see Illus. 6.1):

Step 1. Measure the height and width of the garage door sections.

Step 2. Cut strips of Thermo-Ply of the same dimensions. Thermo-Ply is a tough paper product used as wall sheathing, which comes in 4-foot × 8-foot × ⅛-inch sheets. It has either foil on both sides or foil and white plastic on opposite sides. It is readily cut using only a utility knife, but is very tough.

Step 3. With the door closed, staple or nail the Thermo-Ply panels to the door panels to form aluminum-faced spaces of dead air between the panels. That's it!

NOTE: Thermo-Ply is manufactured by Simplex Industries in Adrian, MI 49221. Consult your lumber dealer for availability in your area. If not available, use ½-inch Thermax insulation the same way.

Bulkhead Door. If you're going to the trouble of insulating your basement walls, then insulate your bulkhead door, too. You probably use the bulkhead door only once in the spring and once in the fall, so the insulation you apply to it can be cheap, awkward, and ugly. Don't forget that the bulkhead door, standing between you and the outdoors, is an *exterior door*. It should therefore be weatherstripped with the same degree of care as your front door. If you never open it, caulk it shut with scraps of fiberglass, and simply install foil-faced fiberglass blankets or batts to the door.

Attic Door. Whether the access to your attic is through a vertical interior door to a stairway or through a scuttle hole in the ceiling, the fact is, the attic door is also an *exterior door*. As we said in Chapter 5, a properly insulated and ventilated attic will run at nearly the outdoor temperature, winter and summer.

The solution is the same as for the bulkhead door: weatherstrip and insulate.

Exterior Door. If your front or back door falls apart, or if you find yourself building an addition with an exterior door, you can build your own insulated exterior door that is simple, low-cost, durable, high-R, and easily planed to fit the existing door frame (see Illus. 6.2). Here's how:

Step 1. Determine the size of the door you want. Although most exterior doors measure 36 inches wide by 80 inches long, this method will allow you to make a door up to 48 inches wide by 96 inches long.

Step 2. Cut two panels of 5/16-inch-thick Luan plywood paneling to the door size.

Step 3. Cut two 2 × 4's to the door's length and three 2 × 6's to the door's width less 7 inches (remember, a 2 × 4 is only 3½ inches wide).

Step 4. Place one of the panels on a perfectly flat, hard surface. Your basement floor may do. Mix up a batch of waterproof glue, such as Resorcinol, apply it liberally to the 2 × 4's and 2 × 6's, and arrange the assembly as shown in Illustration 6.3.

Step 5. Fill the spaces between the wood strips with 1½-inch-thick rigid foam board. Make sure the insulation does not project above the thickness of the strips.

Illustration 6.2
Simple Insulated Door

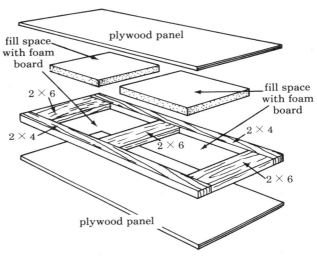

fill space with foam board

plywood panel

2 × 6

fill space with foam board

2 × 4

2 × 4

2 × 6

2 × 4

2 × 6

plywood panel

Step 6. Glue the second panel in place to form a sandwich, then place everything you own on top of this assembly to create uniformly distributed pressure over the 2 × 4's and 2 × 6's. Leave for 48 hours.

Step 7. Apply decorative moldings (available at lumberyards) to the panels as desired, to create a relieved effect. Keep away from the door perimeter, however, which must seal against door stops.

Step 8. Install your homemade door as you would a store-bought door.

Step 9. Seal the door completely with paint or marine varnish.

COST: $40. R-value: 5 to 10, depending on the type of foam.

Windows are nice. They let the sun shine in, allow you to see what the neighbors are doing, and give the kids something to aim at when they're playing baseball. But windows are also burglars. At night, when you're not looking, they steal your heat. A typical insulated home can lose 25 to 30 percent of its heat through windows. If it's a house with a lot of large windows, the figure can climb to 50 percent or higher.

The culprit, of course, is the glass itself. It's a lousy insulator. If you doubt it, here's an experiment you can perform. Make yourself a pot of coffee and pour some of the hot liquid into an ordinary drinking glass. In a minute or so, the outside of the glass will have become so hot you won't be able to pick it up. So much for the insulating ability of glass. Now pour some more of the coffee into a white polystyrene cup. You can nurse the drink for as long as you want and the container won't be warm to the touch, because the foam is an effective insulator.

A single-glazed window, which has an R-value of about 1, conducts 10 to 20 times as much heat from your house as a section of insulated wall the same size. Double-glazed "insulating glass" can halve the loss, but there's still a lot of heat going out the window. You may have already insulated your attic to R-30 or better and your exterior walls to at least R-11. But your windows, at R-1 or R-2, are hanging out like a madman in a T-shirt on a snowy day.

THE OPTIONS

In practical terms, we have two options for reducing the heat loss of windows:
• Add more glazing (either glass or plastic and either inside or outside).
• Add movable insulation that covers the window at night (either a rigid shutter or a flexible shade).

More Glazing. No matter on which side of the house the window is located, double glazing outperforms single glazing. When you add an extra layer of glass, you're creating an air pocket between the panes that inhibits conduction of heat to the outside. Storm windows are the easiest, although not the cheapest, way to add extra glazing. If you already have storm windows, by all means use them. In general, ones with wooden sashes work better than those with aluminum, because wood doesn't conduct heat as easily. Triple-track "combination" storm windows usually offer only minimal insulation because their relatively loose seals allow air infiltration.

Most new storm windows are made of aluminum, and they aren't cheap—about $4 per square foot. Storms eventually pay for themselves in five to ten years.

If you're planning to replace a window entirely, you can buy prefabricated double-glazed windows. These may be simply two separate sheets of glass held together by a weatherproof frame, or two panes fused together. In either case, the seal has to be good, or moisture can build up between the panes, causing condensation and spoiling your view. Permanent double glazing usually is preferable over storm windows for several reasons. The seal around storms is seldom airtight, so some cold air seeps in. Also, there's no time-consuming seasonal mounting and removal with double glazing.

All this talk about extra glass raises a question: If two panes are better than one, aren't three better than two? Maybe. Triple glazing can cut heat loss to two-thirds the loss from a double-glazed window. But it also cuts light transmission by 10 percent. The latter is an important point for south-facing windows, which work for you on a winter's day, letting in heat and light. On north-, east-, and west-facing windows, where solar gain is not a significant consideration, triple glazing performs much better.

Getting that third layer of glazing doesn't come cheaply. Chances are you won't be able to add it yourself to existing windows, so you'll have to buy new, triple-glazed units, which cost about $50 to $60 more per window than double glazing. Unless you're building a new house or doing some serious remodeling, triple glazing may not be worth the investment. If you are building or replacing, you may want to dispense with the third layer of glass on south-facing windows, especially if you have cold but sunny winters.

There is a cheaper way than glass. You can add an extra layer of rigid plastic or plastic film to your existing windows. The plastic is usually placed on the inside of the window. The cheapest of cheap glazings is polyethylene, which costs less than a nickel per square foot. A slightly more expensive option is clear plastic, and the top of the line is rigid plastic. It can be cut to size with a jigsaw or band saw and taped or screwed to your window sash. Although it looks like glass, rigid plastic scratches much more easily and requires some care in handling.

Plastic frame kits are also available. You can buy them at most home centers and hardware stores.

Movable Insulation. Adding extra glazing has certain built-in limitations. At best, you'll be able to boost a window's R-value to 3. More than three layers of glass makes your house noticeably dim inside. If you've set a goal greater than R-3 for your windows, movable insulation is your solution. During the day, you get the full benefit of the glass. At night, when there's not much to see anyway, you can block up the window and stop it from giving away your heat. Movable insulation comes in a variety of forms and materials. You can buy it in kit form, purchase plans, or build it from scratch.

Pop-in shutters—pieces of rigid insulation that are pressed over your windows at night—are a simple, relatively inexpensive form of movable insulation. Kits are available, or you can make your own. Whether you're building or buying, a shutter should be constructed of a lightweight material such as rigid foam board. Anything heavier is difficult to move around and even more of a pain to keep in place.

Which brings us to one of the drawbacks of pop-in shutters: You've got to pop them out each morning, pop them in every night, and find someplace to store them in between. Hinged and sliding shutters are slightly more work to make, but far more convenient in the long run.

Still another way to tuck in your house for the night is with a thermal shade. It operates like an ordinary window shade: You pull it down when you need it and roll it back up when you don't. But unlike your thin vinyl shades, which do little to cut heat losses, thermal shades can reduce heat loss through single-glazed windows by 50 percent or more.

Like shutters, thermal shades can be made from a variety of materials. Thinness is important in insulating shades, because they can get quite bulky when they're rolled up. Shades often incorporate layers of reflective foil separated by air spaces. The shiny material cuts down radiant heat loss by reflecting heat back into the room.

Thermal curtains have gotten a bit of a bum rap, mainly because of false claims made by manufacturers whose products have little or no insulating value. Most of the decorative curtains and drapes on the market aren't designed to stop heat from escaping. They hang loose, away from windows, and let air pass both through and around them. In order to have any value as an insulator, a true thermal curtain must seal at the bottom and sides so cold air gets trapped by it. The material should resist air flow through it, and, preferably, have air spaces or fiber fill for insulation. Like shades, some thermal curtains have foil or silvered backings to reflect radiant heat.

As you see, there are lots of things you can make or buy that will tighten up your windows. But you want to know what works best. In his book, *Movable Insulation* (Emmaus, Pa.: Rodale Press, 1980), William Langdon provides an answer. He did computer simulations of south-, east-, north-, and west-facing windows in Madison, Wisconsin, that show double glazing with R-5 movable insulation to be the winner in a cold climate. Double glazing with insulation is a net energy gainer through the entire season on south, east, and west windows. North-facing windows are incorrigible losers of energy, but the losses with double glazing and insulation were smallest of all the options.

Single glazing with insulation finished second on south-facing windows, but dipped below triple glazing on all others. In every case, single-glazed windows without insulation put in a poor showing as net losers.

If you plan to tighten up your windows a few at a time, those facing north call out for immediate attention, followed by those facing east and west.

46

If you have single-pane windows and can't afford double glazing *and* insulation, or triple glazing, then movable insulation over your single-glazed windows does seem like the next best thing. But expect a slight problem. When insulation is placed over cold, single-glazed windows, the moisture in the room air that's trapped between the two can condense and leave a film of water on the window, and you may have to mop little puddles off your sills and sashes some mornings. Condensation should be much less of a problem if double glazing is used with insulation.

When considering what to do about your windows, you also have to be honest with yourself. Will you be religious about closing the shades and shutters every night, or will you let it go when you're too tired? The saying that "passive solar homes require active owners" is especially true when it comes to movable insulation. The insulation isn't going to do you a whole lot of good if it's buried under a pile of the kids' toys when it should be up in the window. Double and triple glazing, on the other hand, are always on the job, whether or not you think about them.

There are lots of alternatives to throwing your heating dollars out the window. Some are cheap, others are expensive. Some are complex, others incredibly simple. Here are some simple *and* inexpensive solutions.

Inside storm windows. Several companies sell kits for storm windows that install on the inside of windows. These kits are inexpensive, and you don't stand to save much money by assembling the list of materials yourself.

The Insider storm window kit is made by Plaskolite, Inc. (1770 Joyce Ave., Box 1497, Columbus, OH 43216). It has a rigid acrylic glazing and self-adhesive vinyl frames that snap open to let you remove the glazing for the summer. The same company produces the Weatherizer kit containing a thick Mylar film and acrylic tape for attaching the film to your window frame. An Insider kit for a 40 x 80-inch window goes for about $17. Smaller sizes are available for both.

Perkasie Industries (50 E. Spruce St., Perkasie, PA 18944) sells Thermatrol Storm Windows in 30 standard sizes that combine to fit over 350 window sizes. The kits include rigid acrylic glazing in acrylic frames, foam gaskets, and turn locks to hold the windows in place. Two-pane versions allow you to remove the lower pane for ventilation. A 30 × 46-inch two-pane kit costs $22. A 54 × 106-inch two-pane kit is $56. Single-pane kits cost less.

The 3M Company (233-2W 3M Center, St. Paul, MN 55101) offers Flexigard insulating window kits for installation either inside or outside. Flexigard is a clear laminated film that is shatterproof. The frames are aluminum and the windows are held in place by plastic keepers. The cost is $239 for a kit containing 25 yards of four-foot-wide film, 58 six-foot aluminum frame members, foam tape, double-sided tape, and plastic corners and keepers—enough material to cover about 25 normal-sized windows.

Pop-in-shutters. If you want the world's simplest shutter, William Shurcliff, a retired Harvard optics professor, has designed it (see Illus. 7.1). The only drawback is that it requires finding a place to stow it when not installed in the window.

Illustration 7.1
Insulated Pop-in Shutter (Cross Section)

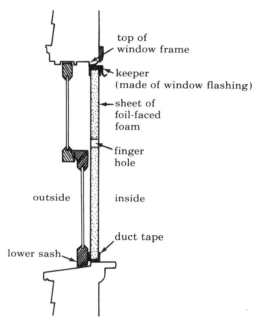

top of
window frame

keeper
(made of window flashing)

sheet of
foil-faced
foam

finger
hole

outside   inside

duct tape

lower sash

Step 1. First determine what surfaces your shutter will contact along the width and height of the window opening.

Step 2. Take a piece of ¾- or 1-inch-thick foil-faced rigid foam board, such as Thermax or High-R Sheathing, and cut it ¼ inch narrower than each of the dimensions taken in Step 1. To cut foam, use a bench saw or sharp utility knife held against a metal straightedge.

Step 3. Tape the exposed edges of the foam to prevent wear. Ordinary duct tape works very well, but other equally suitable varieties in more pleasing colors also are available.

Step 4. Make a keeper to hold the upper edge of the shutter against the top of the window opening by cutting to length a strip of wood or piece of metal, such as aluminum window flashing sold at lumberyards.

Step 5. Place the shutter in the window opening and press it firmly against the lower sash. Then slide the keeper tightly into place against the upper edge of the shutter and mark the keeper's position against the top of the window frame. Remove the shutter and nail or staple the keeper in place. To reinstall the shutter, slip the top edge behind the keeper and press against the bottom edge until it seals against the lower sash.

Step 6. Drill a hole as big around as your finger in the *center* of the shutter for easy removal. The heat loss will be minuscule.

Illustration 7.2
Insulated Bifold Shutter

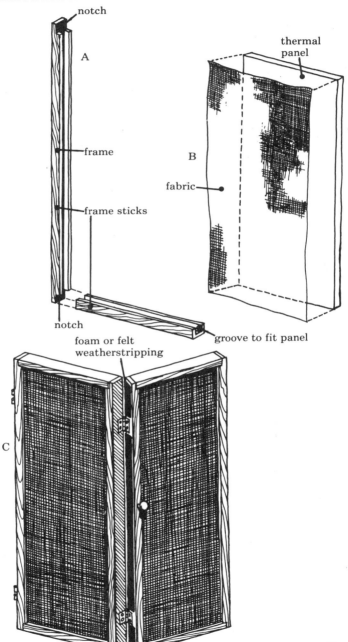

notch

A

thermal
panel

frame

B

fabric

frame sticks

notch

foam or felt
weatherstripping

groove to fit panel

C

49

Fancy bifold shutters. If you are a more advanced handyman and you find the above solutions a little tacky, you can make some really spiffy decorative bifold shutters, which fold to the side of the window when open, thus also solving the storage problem (see Illus. 7.2). These were designed by Jim Eldon, project editor of *New Shelter* magazine.

Step 1. Measure the window opening, then cut wood frame sticks from ¾ × 1⅛-inch #2 pine so that when assembled as shown in the illustration they form the framework for a pair of shutters ¼ inch narrower and ¼ inch shorter than the dimensions of the window opening.

Step 2. Cut a ⅜-inch-deep and ¾-inch-wide groove (dado) in each frame stick and notch (rabbet) the ends of the vertical sticks to accept the ends of the horizontals. Sand and finish, being careful to keep the finish off those areas that will form the joint.

Step 3. Cut pieces of ¾-inch foil-faced Thermax insulation to fit inside the grooves of the frame sticks.

Step 4. Cut the fabric of your choice to the same size as the Thermax and cement the fabric to both faces of each Thermax panel, using an aerosol adhesive.

Step 5. Assemble the frame sticks around each shutter panel with white carpenter's glue and 3d finishing nails.

Step 6. Fasten the pair of shutters together with loose pin hinges. Do not inset (mortise) the hinges, but attach a strip of felt or foam weatherstrip (with cutouts to allow for the hinges) to one of the mating edges.

Step 7. Install another set of hinges to attach the shutters to the window casing. Place a ⅛-inch shim under the shutter while hinging. Apply a weatherstrip filler as in Step 6.

Step 8. Attach a knob to the center of the shutter. This makes opening the shutters easy and keeps the fabric clean.

■Chapter 8■

Make Your House Float: Caulking and Weatherstripping

| THE TIGHTER HOUSE |

It may surprise you, but our living spaces *should* leak a small amount of air. A small amount of infiltrating air is necessary to remove odors, staleness, and chemical pollutants from interior furnishings and building materials, and to prevent the buildup of moisture within the living space. The rate at which fresh air infiltrates and replaces the stale air within a building is measured by the air exchange rate. This is the number of times the complete volume of air within a space changes each hour. In the average home the air changes between one and two times per hour. In a new home the air exchange rate averages once every hour, and in an old house the rate is three, four, or more times every hour. The introduction of outdoor air by infiltration forces the heating and cooling systems to work harder in an effort to stabilize temperatures.

Infiltration is the uncontrolled flow of air into your living space. Ventilation is the controlled and necessary air introduced into a living space. They are equally important in the determination of an optimum air exchange rate and inner comfort. The recognized minimum ventilation standard for respiration and odor control presently is 10 cubic feet of air per minute per person. Converting this figure to the average family of 3½ persons and the average living space of 1,500 square feet, or 12,000 cubic feet, an air exchange rate of ⅕ per hour would satisfy the minimum ventilation requirements.

This is a very difficult rate to achieve. Normal traffic into and out of a home, as well as untraceable leaks, limit the amount we can decrease infiltration. There is also evidence that air exchange rates of less than ½ may be unhealthy, due to the accumulation of air pollutants within the home.

An air exchange rate of ½ seems to be a realistic goal. Illustration 8.1 indicates the potential energy savings in dollars with the reduction of your infiltration rate. For example, a savings of $900 a year is realized when the exchange rate is reduced from 2 to ½ if oil is the heating fuel at $1.50 a gallon.

The two general methods of controlling infiltration are weatherstripping and caulking. Determining which one to use depends on the type of leak.

51

Illustration 8.1
The Cost of Leaks

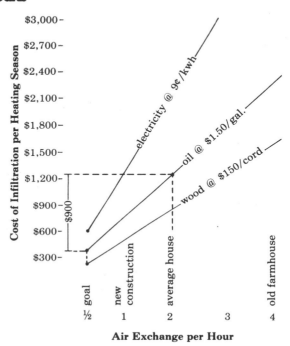

Adapted from *From the Walls In*
by Charles Wing, Atlantic-Little, Brown, 1979.

There are three types of leak to deal with: the fixed joint, the movable or expansion joint, and the operable joint. A fixed joint is one that never moves, such as the frame around a fixed or inoperable window. Movable or expansion joints occur where different building materials meet. Varying rates of expansion between the materials, as a result of fluctuations in temperature or moisture content, can cause gaps in these areas to change in size. Good examples are where horizontal clapboards meet vertical trim boards and where wood meets masonry in a brick home. Doors and operable windows are considered operable joints. *Weatherstripping* is used to reduce infiltration around operable joints, and *caulking* is used for fixed and movable joints.

WEATHERSTRIPPING

Weatherstripping the doors and operable windows in your house could reduce your heating and cooling costs by 10 to 20 percent. However, the choice of the proper type of weatherstripping requires a comparison between products, with special attention paid to the specific use,

Illustration 8.2
Types of Weatherstripping

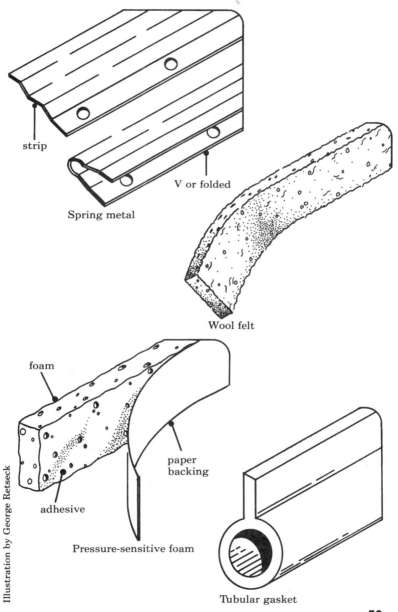

strip

V or folded

Spring metal

Wool felt

foam

paper backing

adhesive

Pressure-sensitive foam

Tubular gasket

Illustration by George Retseck

Table 8.1 Weatherstripping Comparisons

Type	Material	Cost	Quality of Seal	Life Expectancy
Spring metal	brass, bronze, aluminum	high	excellent	20 yrs.
Wool felt	wool	low	good	1-3 yrs.
Pressure-sensitive foam	plastic, vinyl, rubber	low	good	1-3 yrs.
Tubular gasket	plastic, vinyl	moderate	good	2-5 yrs.
Foam-filled tubular gasket	plastic, vinyl	moderate-high	very good	3-6 yrs.
Interlocking metal	aluminum, steel	very high	excellent	20 yrs.
Casement window gasket	plastic, vinyl	moderate	good	10 yrs.

durability, ease of installation, aesthetics, and cost. Illustration 8.2 shows the common types. Table 8.1 compares them.

Spring metal is a flexible metal strip that fills the gaps around doors and windows. It is versatile, very durable, and relatively easy to install. Spring metal comes in two forms: strip and V-metal or folded metal.

Wool felt is an old standby that is very inexpensive, easy to install, and a good performer, but its longevity is reduced considerably by abrasion and cats sharpening their claws. Felt also may not please you aesthetically.

Pressure-sensitive foam, moderately priced, is very easy to install and performs well. However, it can be used only for specific tasks (in situations of compression for example) and does not have a long life expectancy. It also can cause binding problems with doors if not properly sized and installed.

Tubular gasket is a flexible rubber or vinyl tube of weatherstripping generally applied to the exterior of doors and windows. It is an excellent performer, but subject to sun, wind, and rain, and installation is potentially difficult. Painting, in addition, causes gaskets made of vinyl to crack and deteriorate. There is also a variety of tubular gasket which

54

Work with Nonuniform Surface	Ease of Installation	Method of Installation	Notes
no	moderate	nailed	very versatile
no	simple	nailed, stapled, glued	unsightly
yes	simple	glue, self-adhesive	brief lifetime
yes	moderate-difficult	nailed, stapled, glued	exterior application only
yes	moderate-difficult	nailed, stapled, glued	exterior
no	difficult	screwed	difficult installation and maintenance
no	simple	glued, friction	for metal casement windows only

is foam-filled. It performs better and retains both its shape and its effectiveness longer than the regular hollow type.

Interlocking metal (see Illus. 8.3) consists of two separate metal pieces that fit into one another, forming an excellent seal. Unfortunately, interlocking weatherstripping is very expensive, difficult to install, and a potential maintenance problem if dented or bent.

Casement window gasket (not illustrated) is a special type of weatherstripping made to fit snugly around the frame of metal casement windows. It is easy to install, performs well, and is moderately priced.

Choosing the right type of weatherstripping is simple once you realize the types are distinctive and function oriented. The generic type rather than the brand is important, for the quality is competitive. Table 8.1 should help you decide which type of weatherstripping is right for you.

DOOR BOTTOMS

Door bottoms have special weatherstripping problems, such as foot traffic, that require special products. There are four basic types of weatherstripping specifically for this use (see Illus. 8.3):

55

Illustration 8.3
Door Sweeps and Thresholds

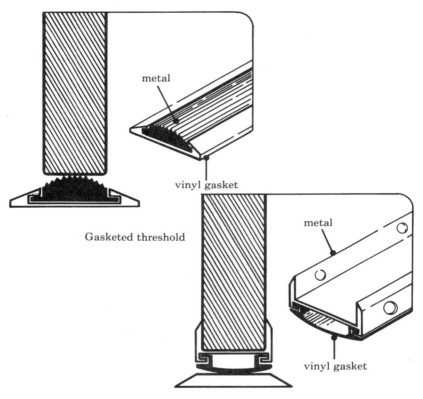

Door shoe

Door sweeps are adjustable strips that are screwed to the bottom face of the door. They should project below the bottom edge of the door enough to make contact with the threshold and create a seal. The door sweep is attached to the inside of the door if the door opens in and on the outside if the door opens out. Door sweeps come in a variety of materials: wool and felt, metal and felt, and metal and vinyl. Sweeps usually don't last very long (one to two years).

Door shoes are foam or vinyl gaskets applied to the bottom edge of the door either by nailing or screwing. They compress and form a seal when the door is closed. Door shoes have a life expectancy of approximately five years and are moderately priced, but require removal of the door from its hinges for installation.

Gasketed threshold is a new or replacement threshold which has a flexible hump beneath the door. The hump compresses when the door is

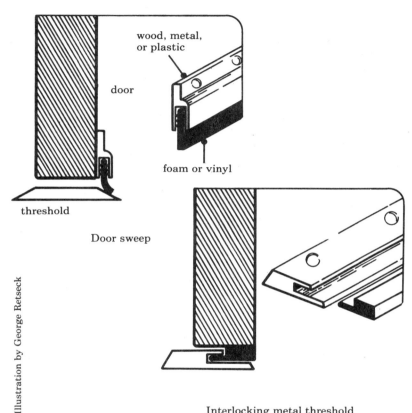

wood, metal,
or plastic

door

foam or vinyl

threshold

Door sweep

Interlocking metal threshold

closed, creating a good seal. The gasket portion of the threshold is generally of a durable vinyl to withstand foot traffic. This type of threshold requires accurate cutting, measuring, and positioning, and proper use of a hack saw.

Interlocking metal thresholds are two-part metal systems: one part as the threshold and the other, similar to a door shoe, that attaches to the door. Although extremely effective, this type of weatherstripping requires a high degree of skill to install and is very expensive.

HOW TO INSTALL WEATHERSTRIPPING

Weatherstripping materials are available at local hardware stores, large catalog outlets, building supply chains and many lumberyards. Most of the materials come in prepackaged kits including nails, screws, etc., and enough weatherstripping to do at least one window or door. Weatherstripping is always installed so the flexible or resilient parts seal out air by making contact. Before installing weatherstripping, make certain the window or door is in good working condition. Repairs, such as tightening hinges and planing door edges, should be done beforehand. Spring metal is our recommendation because of its durability. Here is how to weatherstrip a double-hung window using it. Compared to a double-hung window, weatherstripping an awning or casement window, or a door, is a snap!

Step 1: Four strips of spring metal for the sides should be measured within the window channel to the full height of the sash plus two inches. The other strips—for the top, bottom, and center—should be measured and cut to their exact length only.

Step 2: Open the bottom sash and insert the strips with their flared sides facing out. Slide the strips up until they fit snugly into the channel. Mark positions for nail holes, but before nailing the strips into the channels predrill the holes or use an awl or ice pick to start them. Nail the bottoms of the strips first, then close the window and nail the tops. Nail the middle of the strips last. This "ends first" nailing method is important for keeping the metal strips straight.

Step 3: If the window does not have ropes and pulleys, install the upper channel stripping the same way as you did in Step 2. If the window has pulleys, cut the strips into two pieces and place one above and one below each pulley. Feed the lower portion down the channel and fit it in beneath the rope. As before, nail the strips ends first.

Step 4: Nail the top and bottom strips, which should extend the full width of the window, along the inside edge of the bottom sash and along the top edge of the top sash.

Step 5: Seal the gap between the upper and lower sashes. First check to see if there is room by trying to lock the sash while holding the strip in place. If you can't lock the window, don't install the strip. Your lock probably pulls the top and bottom sashes together tightly enough to obviate weatherstripping anyhow. But if you can install it, attach the spring metal strip to the inner side of the upper sash's bottom rail. Nail the strip so the flared side faces down.

Step 6: Use a wide-bladed tool like a putty knife to pry open the flare a bit, all around the window. This will increase the spring action of the metal.

CAULKING

Caulking compound can be purchased in a number of forms, all of which are easy to find. The type of job determines the form of caulking you should use.

Cartridge. Most caulking compound is available in this form, a cylindrical container with a nozzle for easy application. The cartridge is used in conjunction with a caulking gun, a handy tool which forces the caulking through the nozzle.

Tubes. Small amounts of caulking for special uses are often purchased

58

in hand-held squeeze tubes. Bathtub sealing compounds are often sold in squeeze tubes.

Cans. Caulking compound in cans is available through paint and hardware stores and is used by professionals who use the products in great quantities.

Rope Caulk. Caulk also comes in a roll rather than as a smooth paste, and is very useful for larger cracks.

The cartridge type of caulking compound is the easiest to use and most often proves to be the best buy. It is usually available in two sizes requiring different-sized guns.

Success in sealing leaks is the result of a good application of the proper type of caulking compound. Because many joints are of the movable variety, you'll want to purchase a silicone or neoprene rubber caulk, since these never harden completely, but stay flexible enough to move with the expansion and contraction of the materials that meet at the crack. Such caulks have a lifetime of at least ten years under most conditions.

Many caulking compounds require a primed surface, or special preparation. Instructions often call for warm temperatures. In cold weather, wrap the caulking gun in a heating pad secured with rubber bands. This keeps the caulk flowing and makes it bond better. In extremely hot weather, precool the caulking in the refrigerator.

Caulking will not bond to a dirty surface, so carefully scrape off all old sealant, loose paint, and other debris before you begin sealing a crack. Very dirty surfaces may require cleaning with a solvent.

When applying caulk with a caulking gun, cut the tip of the nozzle at an angle, press the tip against the crack you want to fill, and hold the gun at a 45-degree angle to the crack. While squeezing out the caulk, push the gun away from you. This forces caulk deep into the crack. The line of caulk you lay down is called a "bead." A perfect bead completely fills the crack to a depth at least equal to the crack's width, and does not form too large a ridge above the surface of the crack.

The caulking gun and cartridge type of caulking is very useful with cracks and leaks up to ¼ inch wide; beyond this, special attention is necessary. Large cracks can be treated a number of ways, depending on their location, the type of material, and the seriousness of the crack. If the crack is large enough that insulation can be stuffed into it and covered over, this is the preferred practice. Fiberglass insulation is often used for this purpose, but a more suitable group of foam insulation products is also available. These foams—one such product is called Great Stuff—come in aerosol cans and expand when sprayed, filling gaps nicely. Although expensive, they are very effective and easy to use. Oakum, a treated hemp rope product used since ancient times, is also used to fill large cracks.

Did you ever hear anyone bragging about how well they caulked the crack in their foundation, or how tightly their windows are weatherstripped? Weatherstripping won't impress your friends, but it may cost you less and save more fuel dollars than any other thing you can do.

Chapter 9

The Heat Leak Hit List

THE TIGHTER HOUSE

A lot of research has been done recently on the leakiness of houses. What the researchers are finding is that the obvious leaks (windows and doors), while important, are not as important, even in total, as are all the other little "hidden heat leaks" if they are taken together. Studies have shown that perfect weatherstripping on windows and doors typically reduces infiltration by only 20 to 30 percent.

The researchers (principally those at Princeton University and Lawrence Berkeley Laboratory of the University of California) use infrared cameras and specially adapted blowers placed in entranceways to detect hidden leaks. The infrared cameras detect minute differences in the temperatures of surfaces. By using the blowers to pump up the air pressure inside a house on a cold winter day and viewing the outside of the house or the attic floor with the infrared camera, the researchers can find warm spots—places where warm inside air is escaping. Conversely, by reversing the blower to face outside, lowering the inside pressure and viewing the inside of the house, the camera operator can find cold spots, or spots where cold outside air is infiltrating.

You may be able to find such an infrared service in your area. Insulation contractors sometimes use an infrared camera as a sales tool for insulation. The infrared scan detects areas of missing insulation, but it's even better at finding areas where heat is sneaking *around* the insulation.

Even without an infrared camera, however, you can find cold spots with patience, a small thermometer, and a roll of masking tape on a cold, windy day. Though infrared cameras and thermometers provide hard proof of a heat leak, a little common sense also goes a long way. After all, you don't need an infrared picture to figure out that air is leaking past a clothes-dryer vent stuck in the open position by a ten-year accumulation of lint, or a creosoted fireplace damper with a half-inch gap, or an electrical outlet that blows out matches.

Below is a list of the common hidden leaks researchers have found, where to look for them, and their simple solutions. The numbers correspond to those in Illustration 9.1. Tour your house with the list in hand, note those you have found, and assemble a shopping list of heat-stopping materials. When you're done, you'll be ready to perform the final inspection of the heat loss envelope of your house.

COMMON HEAT LEAK HIDING PLACES

1. The cavity of your exterior wall is probably fairly well ventilated to the outside, and it is the cold air within this cavity that tries to sneak into your house under the baseboards along outside walls. This heat leak can be plugged by cutting thin strips of unfaced fiberglass with a straightedge and a utility knife, then shoving the strips under the baseboards using a putty knife.

2. In modern homes of standard frame construction, interior walls are built with a bottom plate, a top plate, and studs between. The wall frame is assembled on the ground as a unit and then raised into position and nailed. In some older homes, though, interior walls were built in position by nailing the tops of the studs directly to the attic floor joists. The absence of a top plate means a wide-open space from the wall interior up into the attic. Fill such an uncapped wall from above—in the attic—with loose-pour or loose-fill insulation.

3. (Not illustrated) If you live in attached housing such as a row home, town house or "twin," you share at least one wall with a neighbor. This dividing wall, or "party wall" as it is sometimes called, is often hollow. Inside it, little convective loops get going, loops which transfer the coolness of your basement up to your house walls, and the warmth of your house walls up to your attic. The most effective method of blocking this heat loss highway is to blow or pour loose-fill insulation into the wall from the attic until it's filled. Unfortunately, these walls are often constructed of hollow concrete block, which is difficult to fill effectively. At least cap off the top of the walls if they are accessible.

4. At some point in most houses, a furnace flue or chimney penetrates the attic floor and goes up through the roof. The wood framing of the attic floor is boxed out around this penetration, but there is usually a good deal of space in between. This creates an attic bypass heat loss. Heat from the basement and living space rises up into the attic, bypassing the ceiling insulation. To close the gap, simply stuff unfaced fiberglass batt into it. Though you may have reservations about placing fiberglass next to the chimney or flue, fear not. Fiberglass, although not fireproof, won't even char until the temperature reaches 800°F. Your flue or chimney is not likely to exceed 250°F at the attic level, unless something is terribly wrong.

If you use the chimney for wood heat, the best solution is to expose the masonry chimney in the living space below (gaining more heat) and to seal off the gap with ½-inch drywall at the ceiling.

5. Because your attic should be as cold as the outdoors, the trap door or scuttle hole leading to it should be weatherstripped and insulated as described in Chapter 5. If you seldom or never use the trap door, simply insulate it and seal its edges with tape. The trap door is an oft-neglected leakage point, and the consequence of neglect is heat

Illustration 9.1
Hidden Heat Leaks

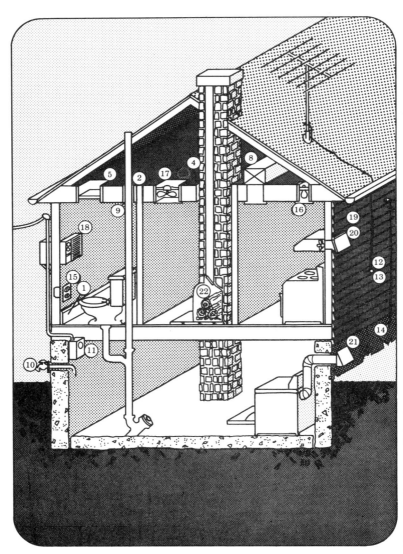

Illustration by George Retseck

loss on the order of leaving a small window wide open.

6. (Not illustrated) If you have a door on your top floor that opens to a stairway to the attic, staple 3½-inch foil-faced fiberglass to the back of the door. Weatherstrip it as well.

7. (Not illustrated) A door to an unheated space, such as a garage, a shed, or a porch, presents the same problem. If the door has no windows and you have no aesthetic objections, insulate it on the unheated side with fiberglass or rigid foam board. Weatherstrip thoroughly.

8. Heating and cooling ducts that travel through the attic deserve special attention because they can be two-time losers. The penetrations they make in the ceiling are cut a little larger than necessary to allow the heating contractor a margin of error. That gap between duct and ceiling should be stuffed with fiberglass. Also, the joints in the duct work should be sealed with duct tape, and the duct then wrapped with R-11 fiberglass blanket or "duct wrap," a vinyl- or foil-faced fiberglass blanket two to four feet wide and one or two inches thick. These measures apply equally to ducts that pass through other unheated spaces, such as basements or crawl spaces.

9. Another penetration through your attic and roof is your plumbing vent. Like the chimney or furnace flue, there is a heat leak around it where it passes from heated to unheated space. Find that spot, stuff it with fiberglass, and seal with duct tape if possible.

10. Outdoor faucets, whether they penetrate the foundation below the sill or an exterior wall above it, leak air. Apply silicone or butyl caulk around the pipe on the outside, and stuff fiberglass into the gap from the inside where the pipe comes through the foundation wall. If the gap is small, simply caulk it.

11. Very close by your fuse box is the spot where a heavy cable comes through the wall, bringing electricity into your house. It may also bring in cold air. Stuff fiberglass into the opening around the cable. At the point on the outside of your house where the cable enters, fill the opening with silicone or butyl caulking. Never try to do this inside the fuse box, however. You risk electric shock, since the power to the heavy cable cannot be turned off except by the utility company.

12. and 13. It's the rare television set that receives its images unaided. Most are hooked up to either an antenna on the roof or a cable for wide-range reception. Follow the cable or antenna wire from the TV set to the place where it first enters the heated space. Stuff the opening with fiberglass on the inside; fill with silicone or butyl caulk outside.

14. The shell of your house has another hole in it to allow a telephone cable into the home. This hole could be almost anywhere, and it's probably difficult to trace the cable from the phone to the hole. So just start looking for a spot—along the sill plate, maybe?—where a thin white cable snakes through, and plug it up with silicone or butyl caulking.

15. This is one of the most crucial items on your hit list. The heat loss headaches caused by electrical service don't stop at the main cable

63

penetration. Your leakage multiplies according to the number of switches and sockets on the inside surface of your outside walls. Behind these switch and outlet plates are boxes, often made of metal, punched full of holes, and 2 inches deep. In houses of standard wood frame construction, these boxes interrupt the vapor barrier, limit the amount of insulation that can be placed between them and the outside wall, and readily conduct heat. On a cold windy day, you will probably be able to feel a draft passing through them.

The way to control these drafts is to install foam rubber gaskets behind the cover plates. These specially made gaskets can be purchased in energy shops, home improvement centers, and hardware stores.

16. If you have recessed lights in your ceiling, consider getting rid of them. As we said in Chapter 5, fire codes insist upon three inches of clearance surrounding the attic floor side of your recessed fixtures. That means you have to get a box or cylinder of plastic or metal to keep attic insulation away from it. And that means a major heat hole. We suggest instead that you simply disconnect, cap, and withdraw the wiring, replace the light with a surface fixture, and insulate to your wallet's content.

17. A ceiling-mounted whole-house fan should be appreciated in summer but suspected in winter. It is yet another interruption of the thermal cap atop your house. You can fix it by making a seasonal plug of rigid board insulation (1½ to 2 inches thick) sandwiched between two pieces of ¼-inch plywood. Cut the plug to fit into the space between the ceiling-level grate and the fan blades. Make it a tight fit, and seal it around the edges with tufts of fiberglass insulation if possible.

18. If you have a room-size air conditioner that you leave in place throughout the winter, you may be letting the wind whistle through its workings. Be sure to caulk all joints where the appliance meets surrounding window or wall, then cover it (inside, outside, or both) with 6-mil polyethylene and seal with duct tape.

19. An even bigger heat hole is found at the exhaust point of kitchen ventilators. Even when you haven't switched on the fan to blow inside air to the outdoors, the passageway often stays open, allowing the same thing at a slower pace. If yours is the always-open type, decide whether you'll really ever need to use it this winter. If not, then cover the opening on the outside wall with polyethylene and duct tape. If possible, stuff an insulating plug in the opening beforehand. If your vent has a mechanism that closes the passageway when not in use, inspect its operation. Clean and replace gaskets if necessary to insure the tightest possible seal.

20. If your bathroom has a similar exhaust fan, there's another vent somewhere to the outside. The vent may be on the wall outside the bathroom or in the attic above it. As with kitchen exhaust systems, consider plugging it for the winter if it's the always-open type. If it closes when not in use, make sure it closes well.

21. Nearly every home has a clothes dryer. Dryers are vented to the outside. Get on your hands and knees and peer up into the outlet.

Surprise! It's stuck in the open position because of lint buildup—right? Buy a better unit with a magnetic closure and *keep it clean.*
22. The open fireplace, beautiful as it is, is a loser. When a blaze is roaring, it takes your warm house air and sends it up the chimney (unless you have a fireplace insert device or tempered glass screen). When there's no fire, a loose-fitting damper will let warm air sneak past it. If you're not going to use your fireplace, fidget with the damper to get as tight a closing as possible, then stuff the cracks with unfaced fiberglass. If you have no damper, cut a board of cement-asbestos or other nonflammable material to cover the flue opening instead. If you do use your fireplace, a board of nonflammable material or a tight-fitting glass fire screen will allow you to cover the hearth opening and go to bed without waiting for the last twinkling embers to die out.

Chapter 10

Hot Water: Waste Not, Sacrifice Not

THE TIGHTER HOUSE

If yours is an average household, your bill for heating hot water is second only to your space-heating bill. The average family consumes about 80 to 100 gallons per day, or over 30,000 fuel-heated gallons per year! Our objective in this chapter is to reduce your hot water bills to the bare minimum without detracting from your life-style.

Common sense dictates that we proceed in the most cost-effective way. That is, we should do those things that return the greatest savings at the smallest cost before considering more exotic options.

A PLAN FOR TOTAL WATER SAVINGS

The average household uses a lot more water than it really needs. In fact, it's often possible for a family to cut its consumption by half and still get the same service and benefits they've always known. Take a look at your water bills. They tell you how many thousands of gallons you use every billing period. With a little division, you can arrive at your water use in gallons per person per day (gpd). You'll probably be shocked.

Listed below are four modifications you can make to your water supply system in order to save water. Savings will vary according to your habits and fixtures, but to illustrate the possible savings of each, included are those achieved by Joe Carter, a book editor at Rodale Press, in his own home.

Pressure-Reducing Valve. This gadget does what it says it does essentially by resisting the flow of pressure from the water main. It costs $40 to $50 and is installed somewhere on the house's water main, downstream from the main shutoff valve but before the line branches off to the water heater or any points of use. To install it, first close the shutoff valve, then cut away a section in the line to make room for the reducing valve. Sweat-solder the valve in place if your pipes are copper; steel pipes require threading.

A pressure gauge at Joe's house showed that the incoming water pressure was 72 pounds per square inch (psi). By installing a pressure-reducing valve, Joe reduced that figure to 36 psi with no inconvenience at all.

Low-Flow Shower Head. A $10 low-flow shower head cut the shower flow rate by 79 percent.

By simply unscrewing the old shower head and installing the new one, a five-minute shower was transformed from a 35-gallon aquatic orgy to a 7½-gallon affair.

Faucet Aerator. A low-flow faucet aerator is essentially a standard aerator with an insert that constricts the passageway to ⅛ inch instead of the normal ½ inch.

Toilet Tank Modifications. Most existing toilets consume 5 to 9 gallons per flush. A mini-flush flapper replaces a standard flapper and saves 1 to 1½ gallons per flush by closing the outflow before the tank is empty. Another way to save is to install tank dams in the toilet tank. These are plastic panels that impound a gallon or two so that only water outside the dam is lost. The ultra-low-cost version of the tank dam consists of putting a couple of bricks in the toilet tank.

Table 10.1 shows the reductions achieved by the modifications installed in various combinations. An overall reduction of 50 percent does not seem an unreasonable goal.

SAVING HOT WATER ALONE

The first step is to tune up the hot water heating system. It's probably wasting more than 30 percent of the energy it consumes. In other words, one out of every three dollars you spend on heating water buys you absolutely nothing. Don't think of solarizing yet. You'd be better off burning dollar bills to heat your water than trying to solarize a system that throws away this much energy.

Fortunately, it's a snap to tune up your hot water system. If you set aside just one day next weekend and work at a comfortable pace with ordinary hand tools, you may reduce your water-heating bills 30 percent. You'll recoup the $80 or so you spend for materials in about a year, and after that, the savings go on forever. Then, with your hot water system operating at peak efficiency, you can go ahead with the final step—a solar water heater.

Temperature Reductions. The fastest and easiest way to improve your water-heating system is to reduce the temperature of the water it produces. Right now your water heater's thermostat is probably set somewhere between 140° and 160°F, and if your home is older, may be set as high as 180°. There's no need for water this hot, because the highest temperature most people can stand for washing and bathing is about 110°F. For almost every other household need, 120°F is sufficient.

Adjusting your water heater's thermostat is as simple as turning a knob or a screw. (You may have to remove an access panel—a 90-second job—to get at the thermostat; and, if the heater is electric, make sure the circuit breaker or fuse is pulled.) Try the "low" or 110°F setting and leave it there for a few days. If you find that you're running out of hot water, or that it's just not hot enough for your needs, turn it back up a few degrees until you've found the lowest possible setting that suits you.

Table 10.1 How Modifications Help

Changes	Psi.	Hot & Cold Water gpd*	Hot Water gpd*	Cold Water gpd*	Percentage of Reduction total hot cold		
1 No modifications	72	46	17	29.7			
2 Pressure reduction only	36	40.6	14	26.5	12 18 11 (compared to #1)		
3 All low-flow devices installed/ high pressure	72	34	14	20	16 0 25 (compared to #2) 26 18 33 (compared to #1)		
4 All low-flow devices installed/ low pressure	36	24	11	13	29 21 35 (compared to #3) 41 21 51 (Compared to #2) 48 35 56 (Compared to #1)		

*Gallons per person per day.

Turning the temperature down from about 150° to about 120° could save you more than $20 a year, which is pretty good return for a three-minute job that costs nothing. Remember: The lower the setting, the more money you save.

Tank Insulation. While the best homes of the 1980s lean toward superinsulation, with R-60 ceilings and R-30 walls, most water heaters are still back in the 1950s. Standard insulation in 75 percent of all new water heaters is still a paltry R-3 for gas heaters and R-6 for electric.

You can easily double or triple the insulation of your water heater for about $20 in materials. If your tank is in a confined area with little open space around it, your best bet is one of the widely available commercial insulation kits (sold in most building supply centers, and plumbing and hardware shops), which use thin, high-R blankets of insulation. Follow the directions that come with the kit.

If you have some space around your heater, you can do even better with a do-it-yourself insulation retrofit, as long as you follow a few simple safety precautions. On gas heaters, leave at least two inches of clear space around the exhaust vent at the top of the tank, and leave a three-inch air space around the base of the unit so the main burner and pilot can get all the air they need. (Some pilots and burners have front-mounted air intakes. Check your owner's manual or call the local dealer for your brand of water heater to locate the air intake of your model.) The insulation must be held in place firmly so it can't settle over the air intakes at some future time.

Illustration 10.1
How to Insulate a Hot Water Tank

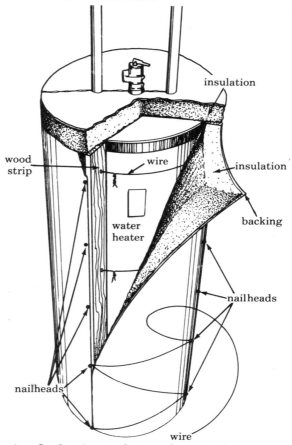

insulation

wood strip

wire

insulation

water heater

backing

nailheads

nailheads

wire

Warning: On electric water heaters, do not insulate over control box or connection to house wiring.

Illustration by Ken Raniere

On both gas and electric heaters, there's also a pressure-relief valve located at the top of the tank, or on nearby plumbing. This valve must be left unobstructed. And, of course, you should leave convenient access to the heater's thermostat and other controls that require periodic attention.

Illustration 10.1 shows how to superinsulate your tank. The specifics depend on the size and shape of your unit, but the general concept is simply to attach vertical wooden strips to the outside of the tank, like fins, and then to staple insulation to the wood. The wooden strips can be cut from any available scrap and should be as wide as the thickness of the insulation you're using (6 inches wide for a 6-inch batt of insulation, for example). You should cut enough strips to divide the outside of the tank into sections the same size as the insulation you're using, usually 16 or 24 inches.

Drill a small hole about an inch from the edge at the top, middle, and bottom of each strip, making the holes in the same location on all the strips. Stand the strips at approximately equal distances around the tank, and thread scrap wire through their holes. Twist the ends of the wires together, tightening them in order to draw the strips against the sides of the tank. As the wires tighten, adjust each strip so that it's resting on its edge, sticking out from the tank like a fin. When all the wooden strips are aligned and the wires are taut, staple foil-backed or kraft-backed fiberglass batts to the wood, completely enclosing the tank except for the controls. Here, instead of stapling the batts, hold the insulation in place with a zigzag wire run around nails partially driven into the adjacent strips. (You can easily remove this zigzag wire to get to the tank's controls.) Finally, cover the top of the tank with heavy batts of insulation, suitably cut out for pipes, vents, valves, and, if a gas heater, the exhaust vent.

This method of insulation is especially good with gas-fired heaters, because the wooden strips keep the insulation from gradually sagging over the air intake and snuffing out your pilot light. If yours is an electric water heater, there is an even simpler method of insulating. Using 6-inch batts of faced fiberglass, cut out a circular piece to fit over the top. Then wrap batts horizontally around the tank, starting at the bottom. Wrap the first batt around the tank so it sits on the floor, and cut its length so it fits snugly but not too tight. Seal the seam with duct tape. Then cut another batt of the same length, wrap it around, and tape the seam as before. The width of the batts (16 or 24 inches) determines how many you need to reach the top. When all the batts are on, seal the horizontal seams with tape and cut out a hole for the control box. A word of warning: Underwriters' Laboratory (UL) has recently cautioned owners of electric water heaters against overinsulating them. Most electric water heaters already contain a layer of insulation between the water tank and the outer shell, and too much insulation can cause enough heat to accumulate within the tank that the wiring and other electrical components can be damaged or cause a fire or shock hazard. Specifically, UL recommends that you do *not* cover over the heater control box, nor should you cover the connection box where the heater is hooked up to the house wiring.

Pipe Insulation. Like tank insulation, pipe insulation can help keep your system's heat in the water, where it belongs. Your local hardware store stocks a variety of self-explanatory do-it-yourself pipe insulation kits that wrap on, zip on, tape on, or snap on. Take your pick. For best results, look for kits offering R-4 or better, and insulate every hot water pipe you can get at. Hot water pipes are easy to identify. If it burns you when you grab it, it should be wrapped. Don't worry about inaccessible plumbing. It's not worth the bother.

Timers. Once your tank is thoroughly insulated, the water inside it will cool off so slowly that you can actually shut off the heater for long periods of time.

Electric heaters can be regulated with an automatic timer, which uses a small clock motor to turn the heating elements on and off at selected intervals during the day. Timers are easy to install (follow the directions that come with the unit), and the only safety consideration is to be sure you buy a model that's rated for the wattage stamped on your heater's identification plate. Gas heaters, which are thermostatically controlled, can be regulated in this manner if you're willing to make a quick trip to the cellar to turn the thermostat to "standby" or "pilot only."

Most families can supply all the hot water they need by running the heater just twice a day, once before and during the morning wash-ups, and again at night from dinnertime to bedtime. Small families may require only 2 to 4 hours of heating a day, while a large family may need 12 hours or more. If you experiment, you almost certainly can find a heating schedule that meets your hot water needs while saving you money.

Timers cost from about $15 to $40, depending on the wattage rating and the number of daily on-off cycles, and they're sold at hardware, plumbing, and electric shops. Annual energy savings from using a timer can easily exceed $20.

Solar Water Heating. Now that you've reduced your use of hot water, you can make better use of solar. Solar domestic water heaters are for real. There are thousands of dealers offering dozens of brands. Fortunately, the shake-down or de-bugging phase is over. Over the last ten years, manufacturers have learned what fluids to use, what sealants stand up, and the critical components to insulate. In fact, most dealers now offer a five-year total (parts plus labor) system warranty.

If you buy a commercial system, get bids from three or more installers. Make sure they are experienced. Ask for the names of past customers. Ask these customers if they are satisfied with the service and the performance they have received. Most problems with solar hot water systems are due to improper installation.

If (and only if) you are handy with plumbing materials and tools, you can build and install a solar water heater yourself. The July/August 1981 issue of *New Shelter* contains complete plans. Cost of the system: less than $1,000.

Regardless of whether you buy or build, you can deduct a full 40 percent of the total installed cost of a solar water heater from your federal income tax. This deduction applies until 1985 and may be spread over several years. Many states allow an additional 5 to 35 percent credit for a total tax credit of 45 to 75 percent. At those rates you can't afford not to go solar.

▰Chapter 11▰▰▰▰▰▰▰▰▰▰▰

Heating Systems: Getting More Degree Days per Gallon

| THE TIGHTER HOUSE |

Most residential heating systems are ripe for improvement. In this chapter we will consider various heating system modifications that may or may not be applicable to your system.

Some of these modifications are gizmos that look strange and have strange names. You have no direct way of judging their true effectiveness. We will discuss these various physical additions and tell you which of them can increase the efficiency of your system.

Illustration 11.1 shows a simplified heating system. The particular system shown is an oil-fired boiler, but the principles and components of oil and gas furnaces and boilers are similar. At the bottom right is the burner. In an oil system, fuel oil from a storage tank is mixed with air and blown as a finely atomized mist into the combustion chamber.

As it exits the burner on its way to the chamber, the fuel and air mixture is ignited by a spark jumping across two electrodes. The rate at which oil is injected is controlled by the size of a nozzle or, more specifically, by the diameter of the hole drilled in the nozzle. The amount of air is controlled by the size of a burner blower air inlet. The ratio of oil and air is adjusted by you or your service man to produce complete combustion—as evidenced by a lack of smoke and a high carbon dioxide concentration (up to 14 percent) in the flue gas going up the chimney. Gas systems are similar except that in older systems the gas escapes from rows of holes drilled into pipes and the gas/air mixture is ignited by a continuous pilot flame. New, more efficient gas systems incorporate two or more of the retrofit options discussed below: power burners, automatic vent dampers, and electric or mechanical intermittent ignition devices.

In both oil and gas systems, combustion occurs in the combustion chamber, a space specially designed to promote complete combustion. Because the walls of the chamber are subject to extremely high temperatures (from 2,000°F to 3,400°F), they are often lined with an insulating refractory material such as firebrick.

Before exiting via the flue pipe and chimney, hot flue gases cross a heat exchanger where much of the heat in the gases is transferred to a

Illustration 11.1
Oil-Fired Boiler

Illustration by Kathi Ember

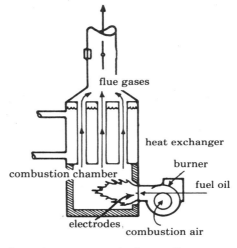

flue gases

heat exchanger

burner

combustion chamber

fuel oil

electrodes

combustion air

fluid that carries the heat from the system to the house. In a warm air furnace the fluid is air; in a forced hot water or "hydronic" system, the fluid is water; and in a steam system it is steam generated by boiling water. Furnaces produce hot air; boilers produce hot water or steam.

Most of the extra cost in a high-efficiency heating system is invested in the heat exchanger. The efficiency of the combustion process is largely determined by the burner design, and the adjustment of the fuel/air mixture is determined by the service man. But the percentage of heat that gets transferred from the flue gases to the working fluid instead of being vented up the chimney is largely determined by the heat exchanger design. Cheaper, less efficient designs are typically fabricated from sheet steel. They have smaller total surface areas for heat exchange, and flue gases make only one simple pass across the system. High-efficiency designs are usually made of cast iron and have large surface areas. Flue gases linger longer in the system, traversing a more tortuous route. In the case of boilers, a "wet-base" design wherein a water jacket surrounds the combustion chamber is more efficient than a "dry-base" design.

After passing through the heat exchanger, flue gas exits via the flue pipe and chimney. The flue pipe usually has a T-shaped fitting with a pivoting disk. This is called the barometric damper. Its function is to keep the fuel/air mixture in the burner constant by introducing air into the chimney as a function of chimney draft. Without the damper to regulate airflow, the greater draft or pull of the chimney caused by lower outdoor temperatures and wind over the chimney would pull excess air into the burner and produce inefficient operation.

Another control found on hot water systems is the boiler aquastat, a water thermostat. Boiler temperatures are maintained during the heating season between upper and lower temperature limits by the aquastat so that heat is always available to the distribution system at the call of the house thermostat. If your domestic (tap) hot water is produced by a tankless coil in your boiler, the boiler must be kept warm through the nonheating season as well. In other words, you're burning oil even in the heat of July. As we will see, reducing the lower temperature limit on the aquastat improves the overall system efficiency.

LOCATING THE LOSSES

Heating system efficiency is defined as the percentage of heat contained in the fuel that ends up in the house, as opposed to the great outdoors. Because of the necessity of venting flue gases to the outside, no fuel-burning heating system is 100 percent efficient. The theoretical peak efficiency imposed by this requirement for oil systems is about 87 percent (13 percent of the flue gas heat is lost up the stack or chimney). However, all systems have further heat losses to varying degrees. Illustration 11.2 shows where the heat typically goes.

**Illustration 11.2
Where the Heat Goes**

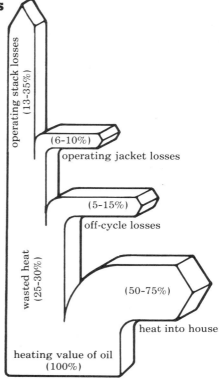

operating stack losses
(13-35%)

(6-10%)
operating jacket losses

(5-15%)
off-cycle losses

(50-75%)

heat into house

wasted heat
(25-30%)

heating value of oil
(100%)

Illustration by Kathi Ember

74

Operating Stack Losses. A properly tuned system with modern burner and high-efficiency heat exchanger loses only 13 to 15 percent of its heat up the chimney while burning. A lower-cost system with an older-design burner, particularly if out of adjustment, can lose 35 percent or more of its heat up the chimney.

Operating Jacket Losses. The jacket is the insulated metal cabinet that houses the combustion chamber and heat exchanger. A well-insulated jacket leaks less than 1 percent of its heat into the surrounding air, while a poorly insulated unit can lose 10 percent or more. Of course, if the heating system is located in a purposely heated space, jacket losses are of no concern because they become heat gains to the house.

Off-Cycle Losses. One of the largest variables in overall heating-season efficiency is the off-cycle loss, i.e., heat lost when the furnace or boiler is not firing. After combustion ceases, air continues to flow into the burner inlet, through the combustion chamber, through the heat exchanger, and up the chimney, thus robbing the system of its stored heat.

THE OPTIONS

There is a bewildering array of adjustments and modifications that can be made to gas and oil heating systems to increase overall efficiency. Whether a modification will result in a significant improvement depends on the deficiencies of the particular heating system involved. An identical modification applied to two different systems may improve efficiency by 10 percent in one but only 3 percent in the other. Furthermore, some of the modifications address the same deficiency in two different ways. You cannot correct a single deficiency twice, so installation of both modifications does not produce twice the savings. Choosing the proper modifications for your system requires an understanding of the deficiencies of your particular system and the ways in which each of the modifications might correct those deficiencies. A proper allocation of your conservation dollars also requires consideration of the payback period of each.

Here's what can be done:

A Tune-Up. If you have an oil-fired heating system, you should clean it at the beginning of every heating season. Many oil dealers provide their customers with service contracts that say the heating system will be cleaned and tuned regularly, but unfortunately, the job only sometimes gets done right. So the cleaning is left up to you, either to do or to supervise. Cleaning the heat exchanger surfaces of accumulated soot (which impairs heat transfer) and adjusting the fuel/air mixture of the burner increase the steady efficiency to the level at which it was designed to perform.

Oil tends to be "dirty" in a combustion process, but natural gas burns cleanly. So annual cleaning and burner adjustment of gas systems are unnecessary. In fact, such measures shouldn't be undertaken for gas-fired systems unless you notice an inexplicable rise in your consumption.

Firing Rate Reduction. An overfiring oil burner is wasting your money. It often is possible to reduce an oil burner firing rate simply by replacing the nozzle with one of a smaller size. The simplest way to determine your present overfiring ratio is to time the burner operation on the coldest night of the year. On that night it should be running continuously. If it runs only 30 minutes out of an hour, it is overfired by a factor of two.

Oil burners frequently have been found to be overfired by a factor of two, but gas burners are often overfired by a factor of 1.5. What's more, it is much more difficult to reduce the firing rate of a gas-fired system. So, tune-ups and firing rate reductions are measures generally limited to oil-fired systems.

Flame Retention Head Oil Burner. In this type of burner, a vaned disk at the burner outlet causes the flame to double back on itself. As a result, incompletely burned fuel is subjected to the flame twice, giving you a higher-temperature flame and a higher steady efficiency. New flame retention head burners also use higher-speed blowers than most older models. As a result, the burner inlet hole is smaller, and the off-cycle loss due to heat-robbing inlet airflow is significantly reduced. Many older burners can be retrofitted with retention heads. If your burner blower is of the slow type (1,750 rpm) you should consider replacing the whole burner.

Gas Power Burner. The gas equivalent to the flame retention head oil burner is the gas power burner. It blows a gas/air mixture into the combustion chamber in exactly the same way as does an oil burner. The benefits are the same increase in steady efficiency and decrease in off-cycle stack losses caused by inlet air flow. Gas power burners are used primarily to convert furnaces and boilers from oil to gas.

Gas Intermittent Ignition. The loss of heat and therefore efficiency due to the constant operation of a gas pilot flame depends on whether the heating system is located in a heated or unheated space, the length of the heating season, and whether the pilot is turned off during the summer. The average pilot light burns gas at the rate of 7,000 cubic feet of gas per year. At a cost of 75¢ per hundred cubic feet (ccf), this represents a cost of $50. Simply turning the pilot off during the nonheating season saves anywhere from almost 7,000 cubic feet in the Deep South to 2,000 cubic feet in the northernmost states. Installation of an electronic or mechanical ignition device further saves from almost nothing in the South to 3,000 cubic feet in the North. Some gas service technicians feel the extra expense of running a gas boiler pilot light during the summer is justified by preventing condensation in and rusting of a boiler. This largely depends on the summertime humidity of your boiler space.

Automatic Vent Damper. The motor-controlled damper (a metal disk blocking the flue pipe) rotates to the open position a few seconds before the burner fires. It remains open until a minute or more after combustion ceases in order to vent any lingering odorous flue gases. (Also, any lingering dangerous flue gases, such as carbon monoxide.) In the closed position, it blocks the flow of air through the flue pipe and thereby reduces off-cycle stack losses. There is a second type of vent damper that closes immediately after combustion, venting lingering gases through perforations in the disk. From an efficiency standpoint, though, the solid type is far superior. A test in one unit showed that holding a solid damper open for three minutes after combustion reduced annual savings by 11.6 percent, but a damper with holes amounting to 8 percent of the disk area produced savings of only 5.5 percent!

It is important to note that the magnitude of vent damper savings is very dependent on heat exchanger design, burner fractional on-time, and the type of burner.

A high-efficiency heat exchanger, a fractional on-time approaching 30 percent, and a retention head burner all increase potential savings by reducing off-cycle stack losses. Table 11.1 shows the damper savings resulting from installation on four different types of systems.

Stack Economizer. As shown in Illustration 11.3, installation of a stack economizer amounts to the addition of another section of heat exchanger. Savings can be dramatic if the existing system heat exchanger is inefficient. On the other hand, a high-quality heat exchanger leaves little further heat to be extracted and will result in very small savings. As a rule of thumb, you shouldn't consider a stack economizer unless your stack temperature is over 500°F.

Some controversy exists over the use of stack economizers because they tend to accumulate soot and lose their effectiveness, and they can reduce flue gas temperature to below 250°F, which in turn can result in corrosive condensation inside the chimney and flue pipe, similar to that produced by creosote and acids found in airtight wood stove chimneys. Both savings and safety hinge upon regular cleaning and maintenance.

Variable Aquastat. As mentioned earlier, a boiler aquastat sets the temperature of the boiler water. Historically, aquastats have been set at temperatures of 180° to 200°F. This gives the distribution-system water plenty of heat-carrying capacity for cold winter nights. Unfortunately, the rate at which a boiler loses heat to its surroundings and up the stack is proportional to the difference in temperature

Table 11.1   Vent Damper Fuel Savings

Burner Design	Boiler Design	Annual Savings in Gallons	Annual Savings in Percent*
Conventional head	Dry-Base Steel	160	11.0
	Wet-Base Cast Iron	100	7.8
Flame retention head	Dry-Base Cast Iron	30	2.3
	Wet-Base Cast Iron	30	2.6

* Savings for New York City
   three-minute time delay damper
   50,000 Btu design load
   40-gal/day domestic hot water
   100% overfired

Illustration 11.3
Stack Economizer

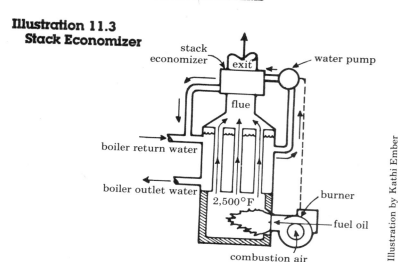

Illustration by Kathi Ember

Table 11.2 Retrofits for Oil and Gas Boilers*

Modifications	Estimated Fuel Saving Range (%)	Approx. Cost ($)	Payback Period (years)
Reduced burner firing rate (by 25%)	6-10[1]	0[2]-45	0-0.3
Boiler water temperature reduction (35°F)	3-7	0[3]-25	0-0.4
Thermostat setback[4] (manual adjustment)	6-10	0	0
Thermostat setback[4] (automatic)	6-10	80	0.3-0.6
Burner efficiency adjustment	2-4	0[2]-45	0-1.0
Retention head burner[5]	6-22[1]	400	0.8-3.2
Vent damper[6]	4-12[1]	300	1.2-3.6
Stack heat reclaimer[6] (economizer)	10-20[1]	450	1.1-2.2
Ducting combustion air from outdoors[7]	0-3[8]	150	2.2 and up
Modern high efficiency burner-boiler	23-25	2,200	4.2-4.6
Blue flame boiler/burner	21-28[1]	2,500	3.5-5.1
Gas power burner	5-10	300	1.5-3.0
Gas intermittent ignition device	0-3	200	4.0 and up
Gas pulse combustion boiler	10-20	2,500	7.5-15

*Savings from refit actions are not additive. Payback period is based on 1,500 gallons per year. Fuel use at $1.50 per gallon or 2,200 ccf (hundred cubic feet) gas at 75¢ per ccf.
1. Based on dry-based steel boiler with nonretention head burner
2. May be included as part of annual servicing
3. Manual adjustment by home owner
4. Setback of 10°F for 8 hours per day
5. Firing rate reduction should accompany burner installation
6. Possible safety hazard exists; long-term testing required
7. Including inlet air damper for burner off-cycle
8. Will vary depending on boiler location in structure

between it and its surroundings. The warmer the boiler, the faster it loses heat. Considering that the water in the distribution system need be only a fraction of the normal 180° to 200°F on the warmer days of fall and spring, you see the value of adjusting the boiler-water temperature in response to outdoor temperature. A variable aquastat can accomplish this automatically. The same results can be obtained manually, as well, simply by turning back the present aquastat control during spring and fall months. Savings are large when the boiler is overfired, but small when the boiler is fired at its proper rate.

Gas Pulse Combustion Boiler. Part of the flue gas heat loss in a normal gas or oil heating system is in the latent heat of water vapor produced as one of the by-products of combustion. The flue gas from oil systems is purposely maintained at above 250°F. If allowed to cool below 212°F, the water vapor would condense as liquid water and combine with sulfur dioxide (also in the flue gas) to form corrosive sulfuric acid in the flue.

Gas combustion, however, produces only water vapor and carbon dioxide. The new gas pulse combustion boiler allows the water vapor in the flue gas to condense and be retained, thus reclaiming the latent heat. The result is a remarkable 93 percent efficiency! Check with your gas company for details on how to obtain these new boilers.

To summarize all this, Table 11.2 shows the range of annual savings and years to payback of the retrofit options listed above, plus several not discussed. It is worthwhile repeating that the savings achieved with combinations of many of the measures are not additive.

Much of the information in this article is derived from Brookhaven National Laboratory, which has been testing oil-fired heating system retrofits for the last two years. A study of gas-fired systems is under way.

▄Chapter 12▄▄▄▄▄▄▄▄▄▄▄▄▄▄▄▄▄▄▄▄▄▄▄▄▄▄

Priorities: Being Your Own Energy Manager

THE TIGHTER HOUSE

The preceding chapters have presented a host of ideas for lowering your energy bills. Some of the ideas are new, some old; some cost a lot, some virtually nothing. Unfortunately, most of us are not in a position to take immediate action on all of the recommendations. For those of us who face life with limited funds, the key question is, "Where should I start, and in what order should I proceed?" This final chapter should help you formulate your own plan of action for saving the most money for the least cost. At the end, we'll give you a couple of warnings about energy savings. With them in mind, you'll see why some people save less than they expect, while others save more.

Calculating the dollar savings of an energy-saving action is complex. It involves the number of heating degree days of the site, the thermostat settings within the house, the human, utility, and solar energy gains, the present and future thermal resistances (R-factors) of the thermal envelope, and the fuel prices and efficiency of the heating system. Table 12.1 lists costs, savings, and other economic factors for 50 common energy-saving actions assuming these conditions: 6,000 heating degree days and an energy cost of $1 per delivered 100,000 Btu's. In other words, the figures in Table 12.1 are based on an F-factor of 1 as represented by the solid-line example in Illustration 12.1.

For example, if a site has 6,000 heating degree days and space-heating energy costs are $1 per 100,000 Btu's, blowing loose-fill insulation into an empty four-inch wood frame wall involves the following factors (see the accompanying box, "Definitions," for a clear understanding of these terms):
• R (present wall thermal resistance) = 4.1
• R (insulated wall thermal resistance) = 13.0
• Cost (per square foot of wall) = $0.90
• First-Year Savings = $0.24
• Benefit (lifetime savings) = $7.20
• Lifetime = 30 years
• Payback = 3.8 years
• Benefit Cost Ratio (BCR) = 8.0

Illustration 12.1
Factoring In Fuel Costs

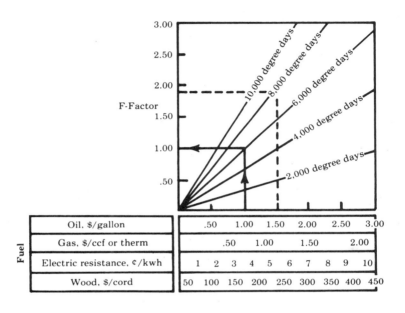

Oil. $/gallon	.50	1.00	1.50	2.00	2.50	3.00
Gas. $/ccf or therm	.50	1.00		1.50	2.00	
Electric resistance. ¢/kwh	1 2 3 4 5 6 7 8 9 10
Wood. $/cord	50 100 150 200 250 300 350 400 450

Fuel

All other heat-loss surfaces are treated in the same way, except doors, windows, and water heaters. For them, unit costs and savings are presented for a door of 20 square feet, a window of 15 square feet, and a 52-gallon water heater.

The two variables in the calculation of savings that vary most widely are degree days and cost per 100,000 Btu's of energy. Table 12.1 can be customized to your situation by multiplying First-Year Savings, Benefit, and BCR by the factor F from Illustration 12.1. Payback should be divided by the same factor. To obtain F, use Illustration 12.1. Locate the point on the lower horizontal scale of the illustration representing the price you currently pay for fuel. Then draw a line vertically to the point of intersection with the appropriate degree day line and horizontally from the intersection to the left-hand scale of F.

In the dotted-line example, suppose you live in an area with 8,000 heating degree days. You heat with electric resistance heaters, paying 5¢/kwh. Find the point 5¢/kwh on the horizontal scale labeled "Electric Resistance," and read up to the sloping line labeled "8,000 degree days." Now, from the intersection read horizontally to the left-hand scale and find F=1.95. Table 12.1 would then be modified by multiplying First-Year Savings, Benefit, and BCR by 1.95. Payback would be divided by 1.95.

Table 12.1 Fifty Common Energy Savers

Heat Loss Area	Conservation Measure	R_o
Basement walls		
1-foot exposure	1½'' Styrofoam, 24'' below grade outside	5.8
1-foot exposure	3½'' studs, fiberglass, drywall inside	5.8
1-foot exposure	1½'' Styrofoam, drywall inside	5.8
4-foot exposure	1½'' Styrofoam, 24'' below grade outside	2.6
4-foot exposure	3½'' studs, fiberglass, drywall inside	2.6
4-foot exposure	1½'' Styrofoam, mastic, drywall inside	2.6
Crawl, 2-foot exposure	6'' fiberglass batts inside	1.7
Floors		
Slab on grade	1½'' Styrofoam to 24'' below grade	12.0
Slab on grade	¾'' urethane, plywood over	12.0
Over unheated basement	6'' unfaced fiberglass & fishline	8.4
1-foot exposure above grade exposed to open air (on piers without skirt)	6'' unfaced fiberglass protected by 1½'' beadboard	3.6
Exterior walls		
Wood frame, uninsulated	4'' loose-fill	4.1
Wood frame, uninsulated	¾'' urethane & ½'' drywall inside	4.1
Wood frame, uninsulated	¾'' urethane & vinyl siding outside	4.1
Wood frame, insulated	¾'' urethane & ½'' drywall inside	11.6
Wood frame, insulated	¾'' urethane & vinyl siding outside	11.6
8'' concrete block	¾'' urethane & ½'' drywall inside	2.6
8'' concrete block	¾'' urethane & vinyl siding outside	2.6
8'' block & brick	¾'' urethane & ½'' drywall inside	3.5
8'' block & brick	¾'' urethane & vinyl siding outside	3.5
Ceilings		
¾'' roof boards	6'' fiberglass, 1'' urethane & ½'' drywall inside	2.2
1½'' wood deck	3'' Styrofoam, ⅛'' plywood & asphalt shingles	3.1
Industrial flat roof	3'' Styrofoam & gravel over uninsulated	3.1
Industrial flat roof	3'' Styrofoam & gravel over 1'' existing fiberglass	7.1
Suspended	R-12 fiberglass 2' × 4' panels	2.2
Doors		
Exterior 3' × 6'8'' wood	aluminum storm door	2.2
Exterior 3' × 6'8'' wood	replace with R-10 thermal door	2.2

R	Cost	First-Year Savings	Benefit	Lifetime	Payback	BCR
12.7	0.85	0.13	3.90	30	6.5	4.6
16.2	1.50	0.16	4.80	30	9.4	3.2
13.8	1.35	0.14	4.20	30	9.6	3.1
9.7	0.85	0.41	12.30	30	2.1	14.5
13.0	1.50	0.44	13.20	30	3.4	8.8
10.6	1.35	0.42	12.60	30	3.2	9.3
20.7	0.35	0.78	23.40	30	0.5	67.0
18.0	0.35	0.04	1.20	30	8.8	3.4
20.0	1.20	0.05	1.50	30	24.0	1.3
25.0	0.50	0.11	3.30	30	4.4	6.6
26.7	0.90	0.35	5.19	15	2.6	5.8
13.0	0.90	0.24	7.20	30	3.8	8.0
10.0	1.70	0.21	6.30	30	8.1	3.7
10.5	1.80	0.21	6.30	30	8.6	2.8
17.5	1.70	0.04	1.20	30	42.0	0.7
18.0	1.80	0.04	1.20	30	45.0	0.7
8.5	1.70	0.38	11.40	30	4.5	6.7
9.0	1.80	0.39	11.70	30	4.6	6.5
9.4	1.70	0.26	7.80	30	6.5	4.6
9.9	1.80	0.27	8.10	30	6.7	4.5
26.5	1.25	0.60	18.00	30	2.1	14.4
18.1	3.00	0.38	7.70	20	7.9	2.6
18.1	2.00	0.38	11.40	30	5.3	5.7
22.1	2.00	0.14	4.20	30	14.3	2.1
14.2	2.00	0.55	16.50	30	3.6	8.3
3.2	125.00	4.09	61.00	15	30.0	0.5
10.0	270.00	10.21	306.00	30	26.0	1.1

Table 12.1 — Continued

Heat Loss Area	Conservation Measure	R_o
Attic		
Floor, no insulation	R-38 loose-fill	3.2
2'' rock wool	increase to R-38 with loose-fill	8.7
6'' fiberglass	increase to R-38 with loose-fill	19.9
Flat overhead	same as attic floor	3.2
Slope	R-19 loose-fill	3.5
Slope	¾'' urethane & ½'' drywall inside	3.5
Crawl space floor	same as attic floor	3.2
Knee wall	6'' blanket or batt	3.9
Windows, 3' × 5' wood		
Single-glazed	aluminum storm window outside	0.9
Single-glazed	acrylic panel inside	0.9
Single-glazed	replace with R-1.9 thermal window	0.9
Single-glazed	R-4 insulating shade or drape	0.9
Single-glazed	R-4 inside insulating shutter	0.9
Double-glazed	aluminum storm window outside	1.9
Double-glazed	acrylic panel inside	1.9
Double-glazed	R-4 insulating shade or drape	1.9
Double-glazed	R-4 inside insulating shutter	1.9
Heating system		
Water heater in heated space	2'' fiberglass kit	3.1
Water heater in unheated space	2'' fiberglass kit	3.1
Hot air ducts in heated space	2'' fiberglass "duct wrap"	1.0
Hot air ducts in unheated basement	2'' fiberglass "duct wrap"	1.0
Hot air ducts in unheated attic	2'' fiberglass "duct wrap"	1.0
Weatherstripping		
Weatherstrip doors & windows	cushion bronze strips	—

HOW TO JUDGE PRIORITY

Given the multitude of energy conservation opportunities surrounding us, which one should we tackle first? Intuition tells many to start with the project having the shortest payback, no matter how small the first-year savings. To others, intuition counsels putting off the small cost/small savings projects until the really big project (with the really big cost) has been accomplished. These two opinions are 180 degrees apart. Obviously, one of them is wrong. In fact, both are wrong! *The greatest total savings will be realized for the lowest total cost by starting with that project having the greatest benefit cost ratio (BCR) and continuing in order of decreasing BCR until the budget is exhausted.*

R	Cost	First-Year Savings	Benefit	Lifetime	Payback	BCR
38.0	0.90	0.41	12.30	30	2.2	13.7
38.0	0.80	0.13	3.90	30	6.2	4.9
38.0	0.70	0.03	0.90	30	23	1.3
38.0	0.90	0.41	12.30	30	2.2	13.7
19.0	0.60	0.34	10.20	30	1.8	17.0
9.4	1.25	0.26	7.80	30	4.8	6.2
38.0	0.90	0.41	12.30	30	2.2	13.7
19.8	0.40	0.30	9.00	30	1.3	22.5
1.9	75.00	12.63	379.00	30	5.9	5.1
1.9	30.00	12.63	126.00	10	2.4	4.2
1.9	130.00	12.63	379.00	30	10.3	2.9
1.8	70.00	12.00	120.00	10	5.8	1.7
1.8	70.00	12.00	240.00	20	5.8	3.4
2.9	75.00	3.92	118.00	30	19.1	1.6
2.9	30.00	3.92	39.00	10	7.7	1.3
3.2	70.00	4.62	46.00	10	15.2	0.7
3.2	70.00	4.62	92.00	20	15.2	1.3
9.4	30.00	10.00	100.00	10	3.0	3.3
9.4	30.00	30.00	300.00	10	1.0	10.0
7.3	1.00	0.00	0.00	20	—	—
7.3	1.00	1.00	20.00	20	1.0	2.0
7.3	1.00	1.33	26.70	20	0.8	26.7
—	.20	.20	2.00	10	1.0	10.0

A project costing more than the remaining budget should be skipped over and returned to as soon as sufficient funds become available.

Intuition #1 (fastest payback) is wrong because it ignores the fact that some improvements have longer lifetimes than others. Table 12.1 provides an example.

Project A: Insulate a water heater in a heated space.

    Cost = $30
    First-Year Savings = $10
    Benefit = $100
    Lifetime = 10 years
    Payback = 3.0 years
    BCR = 3.3

PRIORITIES

Project B: Install a storm window over a single-glazed window.

    Cost = $75
    First-Year Savings = $12.63
    Benefit = $379
    Lifetime = 30 years
    Payback = 5.9 years
    BCR = 5.1

    Project A pays back in only 3.0 years vs. 5.9 years for Project B. But Project B returns $5.10 for each dollar invested vs. $3.30 for Project A. To compare "apples to apples" we must use a common length of time. Over 30 years:

	Project A	Project B
Cost:	90 (3 x $30)	$75
Benefit:	$300 (3 x $100)	$379
BCR:	3.3	5.1

Project B should be our first priority. By this, we see that the

Definitions

    • *Cost.* It is surprising how such a simple concept could be interpreted in so many different ways. Do-it-yourselfers usually mean out-of-pocket expenses only. Using this definition, the "cost" of a barn built of lumber sawed from the owner's trees would be the cost of roofing, fasteners, and gasoline for the chain saw. On the other hand, an economist would recognize the market value of the trees, the value of the owner's labor, and depreciation on the chain saw. An energy audit purports to be a logical analysis of the costs and savings of energy conserving actions. To be logical, it must use the economist's definition of cost. Otherwise there would be no means of distinguishing between two actions producing the same savings, the first requiring but 1 hour of labor and the second 1,000 hours. We will therefore use contracted costs; do-it-yourselfers who disagree can adjust the costs to reflect materials only. All the better for them.

    • *Lifetime.* This is the estimated number of years before the energy-saving action will have to be repeated. At best it is always a guess. A well-maintained storm door may last 20 years or more. The same storm door in a rough neighborhood or a windy location may last only one year. Properly installed attic insulation should last 100 years, but the average building is extensively renovated on a 40-year cycle. We will arbitrarily limit economic lifetimes to 30 years, the common period of mortgages.

    • *First-Year Savings.* This is the amount of reduction in your annual energy bill the first year following the energy-saving action, assuming all other conditions remain the same. One of the limitations of an energy audit is the fact that energy-saving actions interact. Savings could be realized, for example, by insulating the attic, lowering the thermostat, and installing a more efficient requirement to replace an item having a relatively short lifetime can be cost-ineffective, thus making payback an unreliable criterion.

Intuition #2 (do the largest project first) is demonstrated to be wrong by the last entry in Table 12.1: weatherstripping. In fact, weatherstripping consists of dozens of individual tasks that individually cost little and save little, but which in the aggregate can save 10 to 20 percent of the total energy bill.

YOUR GAME PLAN

We will now demonstrate application of the principle of greatest return for minimum investment by formulating a game plan for weatherizing an example house.

Example. The house was built in 1930 in an area having 6,000 heating degree days. It is heated with oil at a cost of $1 per gallon. The basement wall has an exposure above grade averaging one foot. The heating

---

furnace. Should the attic insulation savings be based on the new lower thermostat setting and the higher efficiency of the new heating system, or on the present conditions? Not knowing which, if any, of the other energy-saving actions will be implemented and on what schedule, we have no choice but to treat each action as independent and all other conditions as remaining the same. The home owner should be aware, however, that savings so calculated may be inflated.

• *Benefit.* This is the lifetime savings of the action—the total of the accumulated annual savings produced by the action over its lifetime. Beyond the first year, two factors must be considered: (1) As fuel prices rise, the dollar values of energy savings also rise; (2) As inflation continues, the purchasing power or real dollar value of a saving decreases. If the rate of energy price rise and the rate of general inflation were identical, then the real dollar value (purchasing power) of annual savings would remain the same. We will define benefit in terms of real dollars and assume fuel prices and general inflation keep pace. Benefit therefore equals first-year savings times lifetime.

• *Benefit Cost Ratio (BCR).* This is the benefit of an action divided by its cost. In other words, it is the number of dollars saved or returned to the investor per dollar of initial investment. *It is the only true measure of the priority of an investment.*

• *Payback.* This is the number of years required for savings to equal the initial cost. Using our definition of benefit and the assumption that future fuel prices will increase at the rate of inflation, the years to payback equals cost divided by first-year savings. Payback is not an infallible guide to choosing priorities.

system is forced hot air with ducts that are uninsulated and exposed in the basement. The basement is not currently used as a heated or living space, but the owner plans to finish and heat the area sometime in the future. The electric water heater is located upstairs in the heated space. The following actions are contemplated, but the amount of money available at this time is only $3,200. In what order should the work be done?

Actions:
1. Add 10 inches of cellulose to present 2 inches of rock wool in attic
2. Insulate exterior walls with 4 inches of cellulose
3. Install aluminum storm doors
4. Weatherstrip all doors and windows
5. Insulate (3½-inch fiberglass) and finish basement walls
6. Insulate hot water tank
7. Insulate furnace hot air ducts
8. Install aluminum storm windows

Referring to Illustration 12.1, we find that $F = 1.0$ for heating costs, so the numbers in Table 12.1 do not have to be modified. The single exception is the electric hot water heater. The adjustment factor F for the electric resistance water heater is found from Illustration 12.1 to be 1.5, so we multiply First-Year Savings, Benefit, and BCR by 1.5 and divide Payback by the same factor.

Next, we list the actions and their economic factors in a table (see Table 12.2). Note that action #3, storm doors, fails to pay back within its lifetime. The BCR of 0.5 indicates a return of only 50 cents per dollar invested. We will therefore drop storm doors from the list of actions considered.

Then we assign an economic priority to each action having a BCR of greater than 1.0 and relist the actions in order of priority (see Table 12.3). We also keep a running cumulative cost so that we will not exceed our $3,200 budget. We find that we can accomplish all tasks except finishing the basement wall within the $3,200 limit. It should be noted that insulating the hot air ducts (priority 1) and insulating the basement (priority 7) are contradictory because heat from the ducts in a purposely heated space is not really a loss. Whether to insulate the ducts depends on how long we think it will be before we will have the

Table 12.2

Action	Area or (number)	Cost	First-Year Savings	Benefit	Life-time	Pay-back	BCR	Priority
1. Attic floor	1,152	922	150	4,493	30	6.2	4.9	6
2. Exterior wall	946	851	226	6,811	30	3.8	8.0	3
3. Storm doors	(2)	250	8	122	15	30.0	0.5	0
4. Weatherstripping	1,152	230	230	2,300	10	1.0	10.0	2
5. Basement wall	1,008	1,512	161	4,838	30	9.9	3.2	7
6. Water heater	(1)	30	15	150	10	2.0	5.0	5
7. Hot air duct	100	100	100	2,000	20	1.0	20.0	1
8. Storm windows	(12)	900	152	4,548	30	5.9	5.1	4

Table 12.3

Priority	BCR	Action	Cost	Cumulative Cost	Benefit	Cumulative Benefit
1	20.0	Hot air duct	100	100	2,000	2,000
2	10.0	Weatherstripping	230	330	2,300	4,500
3	8.0	Exterior wall	851	1,181	6,811	11,111
4	5.1	Storm windows	900	2,081	4,548	15,659
5	5.0	Water heater	30	2,111	150	15,809
6	4.9	Attic floor	922	3,032	4,493	20,302
7	3.2	Basement wall	1,512	4,545	4,838	(over limit)

funds to finish the basement walls. Perhaps the solution is to stud and insulate but not finish the walls, thereby saving the cost of insulating the ducts.

By performing priority actions 1 through 6, we find we will save $20,302 with an investment of $3,032. To convince yourself of the principle that proceeding in order of decreasing BCR does indeed result in the greatest savings for the least cost, try proceeding in any other order. By proceeding backwards, for example, starting with priority 7, you'll find that given the limits of the budget, you can accomplish only items 5 through 7 (at a cost of $2,454), and your savings will only amount to $9,481.00.

LAST WORDS

We have attempted to show you the complete spectrum of ways to save energy in your house, along with instructions on how to do it. We have also attempted to tell you, as precisely as possible, how much energy each step will save. There are a couple of warnings you should keep in mind before banking on these estimates, so let's talk about them.

The first warning may clear up some confusion about the cumulative effect of energy-saving measures. We all hear so many claims batted about: cutting infiltration will reduce your heating bill by as much as half; insulating your attic will reduce heat loss by 30 percent; wall insulation will reduce it another 20 percent; basement insulation can drop it another 25 percent. Go after the hidden heat leaks, and you'll get another 20 percent.

Add them all up, and what do you get? A grand total of 145 percent. Very grand and very impossible.

The solution is to stop adding. First of all, not all energy savings are cumulative. Some are interactive. In particular, the more you improve your thermal envelope, the lower your potential savings from heating system improvements. Say, for example, you insulate your attic and walls, thereby reducing your fuel bill from $1,600 to $1,000. Then you add a flame retention head burner to your oil-fired heating system. Instead of saving you 10 percent of $1,600, or $160, the retention head burner saves you 10 percent of $1,000 or $100.

On the other hand, many energy-saving measures are independent of each other. You lose a fixed amount of heat through your walls, whether you have 20 inches of attic insulation, or none. And you lose a fixed amount of heat around wobbly window sashes, no matter what the R-value of the wall alongside them. As a principle, we can say that the effects of insulation are independent of infiltration and that insulating each of the various surfaces of the building envelope is independent of the other. The amount of fuel to be saved is the same, regardless of order.

As a practical matter, we can say this: The quantity of fuel you save by insulating your attic is writ in stone; but the percentage you can save . . . that's writ in sand. Say your house is tidily insulated everywhere but your attic. Adding attic insulation might cut the modest heating bill by 50 percent. Now let's say the entire house is leaky as a sieve. Adding attic insulation can still reduce the fuel bill by the same quantity, but its percentage of the much larger heating bill is proportionately less—around 20 percent.

That's why dealing with percentage savings figures is like chasing a will-o'-the-wisp. Cast a cold eye on all such claims.

Now for the other warning we promised. The biggest question mark in your energy master plan is you. Two families, the same size, can live side by side in two identical houses, and the two household energy budgets can be as different as night and day. Consumption is dramatically affected by the way you operate your house, just as your gas mileage is affected by the way you drive your car. Do you linger around a wide-open front door saying good-bye to guests? Do you use a clothes dryer or a clothesline on a sunny day? Do you smoke enough to need a window opened to clear the air? Reducing total energy consumption requires a life-style audit, too.

Happily, making physical changes to your house seems to make beneficial changes in you. As an example, we'll tell you about a family we know in Maine. We built a sunspace on their house. Our calculations told us the sunspace would reduce their fuel bill by 25 percent. Yet, after a year of operation, they showed us utility bills proving a reduction of 50 percent! What happened? The family, originally novices in the realm of energy, became sophisticated on their own. They were soon fiddling with fans and thermal shutters to get the most solar heat from their sunspace, and became attuned to the energy consequences of other aspects of their lives and household. They became energy-conscious people, and their new awareness was as important in saving energy as the solar addition to their home.

On that note, we'll end. Good luck, and may your bucket not leak!

Index

A

air exchange rate, 51
air leaks
    hiding places of, 61-65
    types of, 52
air locks, 41
Annual Degree Days, 1
aquastat, boiler, 73-74
attic insulation
    installation of, 33-39
    types of, 33-35

B

batt insulation, 5, 22-23, 27-28
bead board, 16
    disadvantages of, 17
    as exterior insulation, 16
benefit, calculating cost for, 87
Benefit Cost Ratio (BCR), 87
blanket insulation, 5, 22-23, 27-28
blown-in foam insulation, 5, 28
boiler aquastat, 73-74
British thermal unit (Btu), 1
bucket analogy, 2-3
bypass areas, in attics, 32-33

C

casement window gasket, 55
caulking
    application of, 59
    forms of, 58-59
condensation, on windows, 47
conduction, 4
Consumer Product Safety
   Commission
    and electrical inspection, 26
    on formaldehyde foam issue, 24
convection, 4
crawl space ventilation (table), 9

D

Degree Day (DD), 1
Design Heat Load, 2
Design Minimum Temperature, 1-2
dew point, 7-8
dollar savings, calculating of, 80-81,
   86-87
doors
    insulation of, 40-43
    weatherstripping for, 57-59
dry rot, and warm temperatures, 7-8

E

energy bills, calculating dollar
   savings and, 80-81, 86-87
energy savers (table), 82-85

exterior insulation
    drawbacks of, 15
    good points, 15
    materials for, 16-17
    steps for, 19-21
    and utility lines, 19

F

faucet aerator, 67
Federal Trade Commission,
   insulation requirements of, 5
fiberglass insulation
    cutting of, 36
    installation of, 22-23, 27-28
first-year savings, and calculating
   costs, 86-87
floor insulation, 29-31
formaldehyde foam, Consumer
   Product Safety Commission on, 24

G

garage door insulation
    installation of, 42
    weatherstripping for, 41-42

H

heat
    common leaks of, 61-65
    flow of, 4
    measuring of, 1-2
heating system
    efficiency of, 74-75
    modifications to, 75-79
    types of, 72-73
heating zone, and R-values
   (table), 34
hot water heating system
    electric, 68-70
    modifications for, 67-71
    solar, 71
hot water tank, insulation for, 68-70

I

infiltration, 50
    control of, 51-52
    and heat loss, 4
    and vapor barriers, 10
insulating window kits, 47
insulation
    exterior, 15-17, 19-22
    installation of, 11, 22-31
    interior, 15-19, 22-23
        attics, 32-39
        floors, 29-31
        walls, 26-29
    methods of, 24-26

Insulation *(continued)*
    movable, 45-46
    and R-values, 5
    types of, 5, 16, 22, 26-30
interior insulation
    effectiveness of, 18-19
    foundation walls (table), 18
    good points, 15-17
    steps for, 22-23
interlocking metal weather-
    stripping, 55

L
leaks, air
    control for, 52
    hiding places of, 61-65
    types of, 52
loose-fill insulation, 5
    blown in from inside, 28
    installation of, 37
loose-pour insulation, 5, 29

M
moisture, 9

P
payback, calculating cost for, 87
polyethylene
    installation of, 11
    as insulation, 10-11
    as window glazing, 45
polyisocyanurate, 16
pressure-reducing valve, 66
pressure-sensitive foam, as
    weatherstripping, 54

R
radiation, and heat loss, 4
relative humidity, 7
retrofits, oil and gas burners
    (table), 78
rigid foam board insulation, 5
    exterior, 29
    installation of, 28-29
    interior, 28-29
    types of, 16-17
R-value
    and absorption of moisture, 16
    and doors, 40-43
    and heating zones (table), 34
    and insulation materials, 5

S
shower head, low-flow, 67
shutters, 47-50
soffit vents, 13
    and insulating precautions, 37
solar water heating, 71
spring metal, as weatherstripping, 54
storm doors, 40-41

National Bureau of Standards
    and, 40
Styrofoam, 16
    as exterior insulation, 16-17

T
temperature, measuring of, 1
thermal conductors, 4
thermal curtains, for insulation, 46
thermal doors, 40
thermal insulators, 4
thermal shades, for insulation, 46
thermographic survey of insulated
    wall, 26
toilet tank modifications, 67
tubular gasket weatherstripping,
    54-55

U
urethane, as exterior insulation, 16
utility lines, and exterior
    insulation, 19

V
vapor barriers, 10-12
    in attics, 32
    as moisture control, 10
    and permeance (table), 8
vent damper fuel savings (table), 77
ventilation, 51
    of crawl space (table), 9
    and moisture, 8
    recommendation for attics, 12
venting system
    and fans, 14
    and roof styles, 12-13
vents, soffit, 13, 37

W
wall insulation, 26-29
water heaters, 68-70
water supply system, modifications
    for, 66-67
water vapor, 10
    and dew point, 7
    and relative humidity, 7
weatherstripping
    to control infiltration, 51-52
    for doors, 55-57
    installation of, 57-58
    types of, 54-55
windows
    condensation on, 47
    glazings for, 44-45
    insulation of, 44-50
    and loss of heat, 44
    shutters for, 47-50
    storm, 45
wool felt, as weatherstripping, 54